Jan M. Weber

Regulation of replication initiation and HML silencing

Jan M. Weber

Regulation of replication initiation and HML silencing

Epigenetics of Saccharomyces cerevisiae

Südwestdeutscher Verlag für Hochschulschriften

Impressum/Imprint (nur für Deutschland/only for Germany)
Bibliografische Information der Deutschen Nationalbibliothek: Die Deutsche Nationalbibliothek verzeichnet diese Publikation in der Deutschen Nationalbibliografie; detaillierte bibliografische Daten sind im Internet über http://dnb.d-nb.de abrufbar.
Alle in diesem Buch genannten Marken und Produktnamen unterliegen warenzeichen-, marken- oder patentrechtlichem Schutz bzw. sind Warenzeichen oder eingetragene Warenzeichen der jeweiligen Inhaber. Die Wiedergabe von Marken, Produktnamen, Gebrauchsnamen, Handelsnamen, Warenbezeichnungen u.s.w. in diesem Werk berechtigt auch ohne besondere Kennzeichnung nicht zu der Annahme, dass solche Namen im Sinne der Warenzeichen- und Markenschutzgesetzgebung als frei zu betrachten wären und daher von jedermann benutzt werden dürften.

Coverbild: www.ingimage.com

Verlag: Südwestdeutscher Verlag für Hochschulschriften GmbH & Co. KG
Heinrich-Böcking-Str. 6-8, 66121 Saarbrücken, Deutschland
Telefon +49 681 37 20 271-1, Telefax +49 681 37 20 271-0
Email: info@svh-verlag.de

Approved by: Essen, Universität Duisburg-Essen, Diss., 2010

Herstellung in Deutschland:
Schaltungsdienst Lange o.H.G., Berlin
Books on Demand GmbH, Norderstedt
Reha GmbH, Saarbrücken
Amazon Distribution GmbH, Leipzig
ISBN: 978-3-8381-3150-4

Imprint (only for USA, GB)
Bibliographic information published by the Deutsche Nationalbibliothek: The Deutsche Nationalbibliothek lists this publication in the Deutsche Nationalbibliografie; detailed bibliographic data are available in the Internet at http://dnb.d-nb.de.
Any brand names and product names mentioned in this book are subject to trademark, brand or patent protection and are trademarks or registered trademarks of their respective holders. The use of brand names, product names, common names, trade names, product descriptions etc. even without a particular marking in this works is in no way to be construed to mean that such names may be regarded as unrestricted in respect of trademark and brand protection legislation and could thus be used by anyone.

Cover image: www.ingimage.com

Publisher: Südwestdeutscher Verlag für Hochschulschriften GmbH & Co. KG
Heinrich-Böcking-Str. 6-8, 66121 Saarbrücken, Germany
Phone +49 681 37 20 271-1, Fax +49 681 37 20 271-0
Email: info@svh-verlag.de

Printed in the U.S.A.
Printed in the U.K. by (see last page)
ISBN: 978-3-8381-3150-4

Copyright © 2012 by the author and Südwestdeutscher Verlag für Hochschulschriften GmbH & Co. KG and licensors
All rights reserved. Saarbrücken 2012

Abstract

The *Saccharomyces cerevisiae* origin recognition complex (ORC) displays a dual role in silencing and initiation of DNA replication, but requires the contribution of auxiliary factors such as Rap1 or Sum1 for full initiation capacity. In this study, the influence of ORC and factors binding in the vicinity of ORC on origin activity and *HML* silencing was analysed.

The silent mating-type loci *HML* and *HMR* of *S. cerevisiae* contain mating-type information that is permanently repressed. This silencing is mediated by flanking sequence elements, the E- and I-silencers. They contain combinations of binding sites for Rap1, Abf1 and Sum1 as well as for ORC. Together, they recruit other silencing factors, foremost the repressive Sir2/Sir3/Sir4 complex, to establish heterochromatin-like structures at the *HM* loci. However, the *HM* silencers exhibit considerable functional redundancy, which has hampered the identification of further silencing factors. Therefore, a synthetic, minimal *HML*-E silencer (*HML-SS* ΔI) that lacked this redundancy was constructed during the course of this study. It consisted solely of Rap1 and ORC binding sites and the D2 element, a Sum1 binding site, and all three elements were crucial for minimal *HML* silencing. This silencer was sensitive to a mutation in *RAP1*, *rap1-12*, but less sensitive to *orc* mutations or *sum1*Δ. Moreover, deletions of *SIR1* and *DOT1* led to complete derepression of the *HML-SS* ΔI silencer. This fully functional, minimal *HML*-E silencer will therefore be useful to identify novel factors involved in *HML* silencing.

Replication initiation at origins of replication in the yeast genome takes place on chromatin as a template, raising the question how histone modifications, for instance histone acetylation, influence origin firing. Initiation requires binding of ORC to a consensus sequence within origins and of other proteins, for instance Sum1, to recognition sites in the vicinity of ORC to support initiation. Sum1 is part of the Sum1/Rfm1/Hst1 complex that represses meiotic genes during vegetative growth via histone deacetylation by the histone deacetylase (HDAC) Hst1.

In this study, it was found that Sum1 functioned in initiation as a component of this complex, implying a role for histone deacetylation in origin activity. Several origins were identified in the yeast genome whose activity depended on both Sum1 and Hst1. Importantly, *sum1*Δ or *hst1*Δ caused a significant increase in histone H4 lysine 5 (H4 K5) acetylation levels, but not other H4 acetylation sites, at those origins. Furthermore, mutation of lysines to glutamines in the H4 tail, which imitates the constantly acetylated state, resulted in a reduction of origin activity comparable to that in the absence of Hst1, showing that deacetylation of H4 was important for full initiation capacity of these origins.

Zusammenfassung

Der *origin recognition complex* (ORC) von *Saccharomyces cerevisiae* spielt eine bedeutsame Rolle bei den Prozessen des *silencings* und der Initiation der DNA Replikation. Zudem werden Hilfsfaktoren wie Rap1 oder Sum1 für die volle Initiationsaktivität benötigt. Im Rahmen dieser Arbeit wurde der Einfluß von ORC und anderen Faktoren, welche in der Nähe von ORC binden, auf die Replikationsinitiation und auf das *HML silencing* untersucht.

Die stillen Paarungstyploci *HML* und *HMR* von *S. cerevisiae* beinhalten permanent reprimierte Paarungstypinformationen. Dieses *silencing* wird durch flankierende Sequenzelemente vermittelt, die E- und I-*silencer*, die über Bindestellen für Rap1, Abf1, Sum1 und ORC verfügen. Gemeinsam rekrutieren sie andere *silencing*-Faktoren, insbesondere den Sir2/Sir3/Sir4 Komplex, um heterochromatinartige Strukturen an den *HM* Loci zu etablieren. Jedoch weisen die *HM silencer* ein hohes Maß an funktioneller Redundanz auf, welche die Identifikation anderer *silencing*-Faktoren bisher verhinderte. Aus diesem Grund wurde im Rahmen des Projekts ein synthetischer *HML*-E *silencer* (*HML-SS* ΔI) hergestellt, dem diese Redundanz fehlt. Dieser bestand nur aus Bindestellen für Rap1, ORC und dem D2 Element, der Sum1 Bindestelle, welche alle essentiell für minimales *HML silencing* waren. Der *silencer* war sensitiv für Mutationen in *RAP1*, *rap1-12*, aber weniger für *orc* Mutationen oder *sum1*Δ. Ferner führten Deletionen von *SIR1* und *DOT1* zu kompletter Dereprimierung des *HML-SS* ΔI *silencers*. Dieser voll funktionelle, minimale *HML*-E *silencer* wird daher für die Identifikation neuer Faktoren, die beim *HML silencing* involviert sind, nützlich sein.

Replikationsursprünge sind im Hefegenom in Chromatin verpackt. Dies wirft die Frage auf, auf welche Weise Histonmodifikationen wie Acetylierungen die Replikation beeinflussen. Für die Initiation wird die Bindung von ORC an eine Konsensussequenz innerhalb des Replikationsursprunges und darüber hinaus die weiterer Proteine wie Sum1, in der Nähe von ORC benötigt. Sum1 ist Bestandteil des Sum1/Rfm1/Hst1 Komplexes, der während des vegetativen Wachstums mittels Histon-Deacetylierung durch die Histon-Deacetylase (HDAC) Hst1 meiotische Gene reprimiert.

In der Studie wurden mehrere Replikationsursprünge im Hefegenom identifiziert, deren Aktivität von Sum1 und Hst1 abhängt. *sum1*Δ und *hst1*Δ bewirkten hier spezifisch am Histone H4 Lysin 5 einen signifikanten Acetylierungsanstieg. Mutationen von Lysin zu Glutamin im H4-Rest, die eine ständige Acetylierung simulieren, resultierten in verminderter Replikationinitiation wie im Falle von *hst1*Δ. Dies bestätigte, dass die Deacetylierung des Histons H4 für die volle Initiationsfähigkeit an diesen Replikationsursprüngen nötig ist.

Contents

Abstract 1

Zusammenfassung 2

List of figures 7

List of tables 8

Abbreviations 9

1. Introduction 11

1.1 From early genetics to the post-genomic era 11

1.2 Epigenetics 11

1.3 DNA compaction and chromatin structure 12

1.4 Nucleosome remodelling and histone modifications 16

    1.4.1 Remodelling complexes 16

    1.4.2 Histone modifications 17

1.5 Heterochromatic regions in *S. cerevisiae* 20

    1.5.1 Silencing at the silent mating type loci *(HM)* 20

    1.5.2 Telomeric and rDNA silencing 23

1.6 Replication in *Saccharomyces cerevisiae* 24

    1.6.1 Replication Origins 24

    1.6.2 Replication origins in other eukaryotes 26

    1.6.3 Regulation of initiation function 27

1.7 The correlation between silencing and replication initiation 28

    1.7.1 Dual role of ORC 28

1.7.2 Sum1 in *HM* silencing and replication initiation 29

1.7.3 Influence of histone modifications on origin activity 29

1.8 Outline of this thesis 31

2 Methods 32

2.1 *E. coli* strain 32

2.2 *E. coli* growth conditions 32

2.3 *Saccharomyces cerevisiae* media and growth conditions 32

2.4 *S. cerevisiae* strain construction 32

2.4.1 Gene disruption 32

2.4.2 Chromosomal integrations 33

2.4.3 Crossing, sporuation and tetrad dissection of *S. cerevisiae* strains 33

2.5 Plasmid constructions 38

2.5.1 ARS fragments 38

2.5.2 *HML*-E fragments 39

2.6 Plasmid maintenance assay 41

2.7 Antibodies 42

2.8 Chromatin immunoprecipitation (ChIP) 42

2.9 Quantitative real-time PCR 43

2.10 Silencing assays 43

2.10.1 Telomeric silencing assay 43

2.10.2 *HML* silencing assay 44

2.10.3 Deletion library *HML* silencing screen 44

2.10.4 Deletion library transformation with synthetic *HML*-E plasmid 44

3 Results 45

3.1 A novel minimal *HML*-E silencer caused *HML* derepression in *sir1*Δ and *dot1*Δ strains 45

3.1.1 Three *HML*-E core elements were sufficient to establish *HML* silencing 45

3.1.2 The Rap1 and ORC binding sites and the D2 element were essential for *HML* silencing 47

3.1.3 Mutations *in trans* caused a reduction in silencing by the minimal *HML*-E silencer 48

3.1.4 Sir1 and Dot1 were required for silencing of *HML-SS* ΔI 50

3.1.5 Genetic screen to search for additional novel factors influencing *HML* silencing 51

3.2 Histone deacetylation affected replication initiation at a subset of origins 61

3.2.1 *hst1*Δ and *rfm1*Δ were synthetically lethal with an *orc2-1* mutation 61

3.2.2 ARS activity of selected origins depended on Sum1 and Hst1 62

3.2.3 *sum1*Δ and *hst1*Δ caused increased histone H4 aceylation at selected replication origins 66

3.2.4 Changes in H4 acetylation caused defects in plasmid stability 67

3.2.5 Histone H4 mutations did not cause synthetic lethality with *orc2-1* 68

3.3 *sum1*Δ caused synthetic growth defects in combination with mutations in sister chromatid cohesion factors 70

3.4 Factors that bind in proximity to ORC binding sites were essential for viability of *orc2-1* mutant strains 72

3.4.1 *dat1*Δ, *gat3*Δ and *rgm1*Δ were synthetically lethal with an *orc2-1* mutation ... 73

    3.4.2 *dat1Δ* and *gat3Δ* did not influence origin activity on chromosome X 76

    3.4.3 *dat1Δ* and *gat3Δ* did not influence silencing of the telomeres or the *HML* locus 77

4. Discussion 82

    4.1 A synthetic *HML*-E construct was sufficient for *HML* silencing 82

    4.2 *HML-SS ΔI* was sensitive for some, but not all previously identified mutations that influence *HML* silencing 85

    4.3 High-throughput genetic screens as useful methods to identify novel silencing factors? 86

    4.4 Function of the Sum1/Rfm1/Hst1 deacetylation complex in replication initiation 90

    4.5 Did Sum1 affect sister chromatid cohesion? 92

    4.6 Identification of additional factors with DNA-binding proximal to ORC 93

    4.7 Summary of the main results of this study 95

References 96

Acknowledgements 109

Publications 110

List of figures

Fig. 1 Illustration of DNA compaction levels in eukaryotes — 13
Fig. 2 Schematic representation of the nucleosome structure — 14
Fig. 3 Illustration of three different models for the arrangement of the 30-nm fibre — 14
Fig. 4 Models for the formation of the condensed chromatin structure — 15
Fig. 5 Illustration of the conversion of chromatin states by N-terminal modifications — 18
Fig. 6 Representation of known N-terminal histone H4 modification in *S. cerevisiae* — 19
Fig. 7 Schematic representation of the *S. cerevisiae HM* loci — 21
Fig. 8 Model for heterochromatin formation at *HML* — 22
Fig. 9 Schematic representation of the process of replication initiation — 26
Fig. 10 Model for histone deacetylation by the Sum1/Rfm1/Hst1 complex — 30
Fig. 11 Design of a synthetic *HML*-E silencer — 46
Fig. 12 Characterization of synthetic *HML*-E silencer constructs — 48
Fig. 13 Synthetic *HML*-E was sensitized for mutations in *RAP1* and ORC and for the deletion of *SUM1* — 49
Fig. 14 *sum1*Δ and *orc2-1* increased *HML-SS* ΔI derepression in *mat*Δ cells — 50
Fig. 15 Sir1 and Dot1 were essential for silencing of *HML-SS* ΔI — 51
Fig. 16 Illustration of the experimental procedure to identify novel factors that influence silencing by the synthetic *HML*-E silencer — 52
Fig. 17 Re-evaluation of *HML-SS* ΔI defects in candidate strains — 53
Fig. 18 Secondary screen of selected candidate strains for *HML-SS* ΔI silencing defects — 53
Fig. 20 Variability of silencing defects in genetically identical segregants — 57
Fig. 21 *hir2*Δ did not affect silencing of *HML-SS* ΔI — 58
Fig. 22 *mrpl20*Δ did not cause *HML-SS* ΔI derepression — 59
Fig. 23 HIR complex components did not influence *HML-SS* ΔI silencing — 60
Fig. 24 *hst1*Δ and *rfm1*Δ showed synthetic lethality with *orc2-1* — 61
Fig. 25 Search for intergenic regions that bind both Sum1 and ORC — 62
Fig. 26 Sum1 and Hst1 were necessarry for ARS activity of selected origins — 64
Fig. 27 Schematic representation of the ARS sequences analyzed in this study — 65
Fig. 28 *sum1*Δ and *hst1*Δ caused increased acetylation of H4 K5 at selected origins of replication — 66
Fig. 29 Mutation of lysine to glutamine in the N-terminal tail of histone H4 caused a similar initiation defect as *sum1*Δ and *hst1*Δ — 68

Fig. 30 Mutation of lysine to glutamine in the N-terminal tail of histone H4 caused a minor growth retardation in *orc2-1* cells ... 69
Fig. 31 *sum1Δ* showed a slight synthetic growth defect with the sister chromatid cohesion mutant *smc3-42* ... 71
Fig. 32 *sum1Δ* showed a slight growth defect with *ctf18Δ* ... 72
Fig. 33 Search for intergenic regions that bind both Dat1 and ORC ... 73
Fig. 34 Search for intergenic regions that bind both Gat3 and ORC ... 74
Fig. 35 *dat1Δ*, *gat3Δ* and *rgm1Δ* showed synthetic lethality with *orc2-1* ... 76
Fig. 36 Dat1 and Gat3 were dispensable for ARS activity of origins from chromosome X ... 77
Fig. 37 *dat1Δ* and *gat3Δ* did not affect telomeric *URA3* silencing ... 78
Fig. 38 *dat1Δ* and *gat3Δ* caused a minor reduction in telomeric *ADE2* silencing ... 79
Fig. 39 *dat1Δ* and *gat3Δ* did not affect silencing of natural telomeric *URA3* insertions ... 79

List of tables

Table 1: *Saccharomyces cerevisiae* strains used in this study ... 34
Table 2: Oligonucleotides used for knock-outs and molecular cloning ... 36
Table 3: Oligonucleotides used for generating synthetic *HML*-E fragments ... 39
Table 4: Combination of oligonucleotides used to generate synthetic *HML*-E fragments ... 40
Table 5: Plasmids used in this study ... 40
Table 6: Oligonucleotides used for the construction of ARS plasmids used in plasmid maintenance assays ... 42
Table 7: Oligonucleotides used for quantitative real-time PCR of ChIP samples ... 43
Table 8: Complete list of secondary screen candidates ... 55
Tables 9: Top list of gene disruptions with reduced mating ability in a genetic screen re-array ... 56
Tables 10: Top list of gene disruptions with reduced mating ability with *HML-SS* ΔI plasmid ... 56
Table 11: List of additionally tested secondary candidates ... 56
Table 12: Gene expression change in *sum1Δ* and *hst1Δ* strains compared to wild-type ... 62
Table 13: Intergenic regions with high co-occurrence of Dat1 and ORC binding ... 74
Table 14: Intergenic regions with high co-occurrence of Gat3 and ORC binding ... 75
Table 15: Intergenic regions on chromosome X with high co-occurrences ... 77
Table 16: List of factors with high probability of binding proximal to ORC ... 80

Abbreviations

5-FOA	5-fluoro-orotic acid
ACS	ARS consensus sequence
ARS	Autonomously replicating sequence
ATP	Adenosintriphosphat
bp	base pair
ChIP	Chromatin Immunoprecipitation
DMSO	Dimethylsulfoxid
DNA	Deoxyribonucleic acid
H_2O	Distilled water
HAT	Histone acetyltransferase
HDAC	Histone deacetylase
HDM	Histone demethylase
HM	Homothallic mating
HML	Homothallic mating left
HMR	Homothallic mating right
HMT	Histone methyltransferase
Hst	Homologue of Sir two
kB	kilo bases
LB	Luria-Bertani medium
MAT	mating type locus
MCM	Minichromosome maintenance
NAD	Nicotine adenine dinucleotide
OD	optical density
ORC	Origin recognition complex
ORF	Open reading frame
PCR	polymerase chain reaction
Rap	Repressor activator protein
Rfm	Repression factor of MSEs
RNA	Ribonucleic acid
SDS	Sodium dodecyl sulfate
Sir	Silent information regulator
Sum	Suppressor of mar
wt	Wild-type

YM	Yeast minimal medium
YPD	Yeast peptone dextrose medium
YOGRT	Yeast-based, oligonucleotide-mediated gap repair technique

Yeast genes are named according to the *Saccharomyces cerevisiae* genome database (SGD) gene nomenclature conventions:

http://www.yeastgenome.org/help/yeastGeneNomenclature.shtml

For amino acids, the one letter code was used, for instance: K = lysine; Q = glutamine

1. Introduction

1.1 From early genetics to the post-genomic era

The mechanisms of inheritance have been extensively studied in many classical genetic experiments since the systematic analyses by Gregor Mendel almost 150 years ago. After the discovery of desoxyribonucleic acid as carrier of the genetic information (the "transforming principle" (Avery *et al.*, 1944)), the triplet code, which underlies the protein synthesis, was identified in the 1960's (Crick *et al.*, 1961). The release of the first eukaryotic genome 1996 (*S. cerevisiae*) marked a key step in the biological field of genetics that was followed by whole genome sequencing of several other model organisms at the end of the 20th century. In 2001, eleven years after its foundation, the Human Genome Project reached a milestone with the publication of the sequence of the human genome (Lander *et al.*, 2001). Surprisingly, neither the amount of nucleotides ($\sim 3.2 \times 10^9$ bp) nor the number of genes (20,000 – 25,000) was exceptionally high compared to other organisms. Furthermore, the genome-wide nucleotide divergence between human and chimpanzee was determined to be only 1.23% (The Chimpanzee Sequencing and Analysis Consortium, 2005). With $\sim$ 100,000 genes, the wild cabbage (*Brassica oleracea*) genome contains about four times more genes than the human genome, although it is smaller (6 - 8.7 x 10^8 bp). To date, the largest genome (11-38 times bigger than the human genome) is estimated for the marbled lungfish *Protopterus aethiopicus* (Hallstrom & Janke, 2009). These data indicate that not the nucleotide sequence length or gene number alone, but the precise control of the gene expression is crucial for the development of higher organisms. To solve the questions concerning this regulation process will be the task of the post-genomic era and is addressed by the field of epigenetics.

1.2 Epigenetics

After the complete human genome sequence had been determined, a follow-up project named ENCODE (ENCyclopedia Of DNA Elements) was launched in the year 2003 (ENCODE, 2004). The aim of this project was to identify all functional elements within the human genome, among these (besides the actual genes) all repressors and silencers, origins of replication and chromatin modifications. As conserved processes are best analysed first in less complex model organisms, the single-cell eukaryote *Saccharomyces cerevisiae* is an excellent object for studying the mechanisms that control gene expression. Except for so-called housekeeping genes, which are always expressed, the majority of genes is subject to gene regulation. During cell-cycle progression or depending on the developmental stage, environmental conditions or tissue of higher, multicellular organisms, specific genes are switched on, while others that are not required under the given circumstances at that time, are switched off. To accomplish this, several regulatory mechanisms have evolved. A broad range of sequence-specific DNA-binding proteins can activate (transcription factors) or repress (transcriptional repressors) transcription in a dynamic manner. Beyond that, daughter cells are capable of inheriting the gene expression profile from their mother cell. Although the genomic sequence does not change, varying transcriptional states due to modulated DNA-protein interactions can be passed from one cell to the other − a biological phenomenon termed "epigenetics". The functional targets of the ENCODE identification process mentioned above all play an important role in epigenetic mechanisms and will be elucidated in the following, separate sections. In contrast to promoter-specific transcriptional inactivation, epigenetic gene repression via a process termed "silencing" is gene-independent and affects broad chromosomal regions. This is achieved by chromatin architecture of the eukaryotic chromosomes, which controls the accessibility of the DNA double helix to the transcription machinery.

1.3 DNA compaction and chromatin structure

Unlike the prokaryotic genome, which is organised in a ring-like structure in the cytosolic nucleoid, the eukaryotic DNA is highly condensed within the cell nucleus. Several steps of compaction can be distinguished, starting with the DNA double helix with a width of 2 nm to the densest packaging of the chromosomes with a diameter of ~ 1400 nm (Fig. 1).

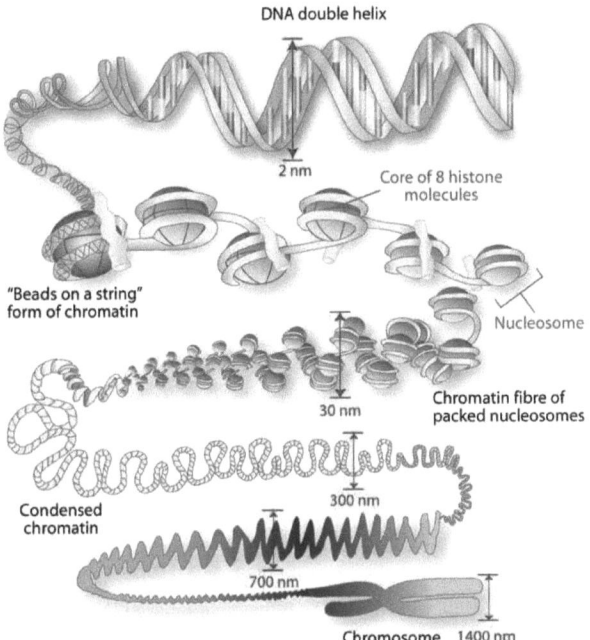

Fig. 1 Illustration of DNA compaction levels in eukaryotes
The packaging from the 2-nm diameter DNA double helix to a eukaryotic chromosome with an extension of 1,400 nm is shown. Compaction levels in between comprise the "beads on a string" chromatin structure built by the DNA strand wrapped around a histone octamere and thereby forming nucleosomes and the more compact 30-nm fibre, which in turn is condensed. The picture is taken from the EPIgenetic TReatment Of Neoplastic disease project homepage (www.epitron.eu).

The so-called "30-nm fibre" is made by several levels of folding of the chromatin organisation ("beads on a string"), which is facilitated and stabilized with the help of proteins like the mammalian linker protein histone H1 (Contreras et al., 2003). Chromatin consists of core units, the nucleosomes, that are wrapped ~ 1.7 times with ~ 147 base pairs (bp) of DNA and connected with short linker sequences (~ 20 bp) making about 75-90 % of the genomic DNA wrapping nucleosomes (Richmond & Davey, 2003). The nucleosome is built of a histone octamere of each two molecules of histone H2A, H2B, H3 and H4 (Luger et al., 1999), which contain N-terminal histone tails (Fig. 2).

1. INTRODUCTION

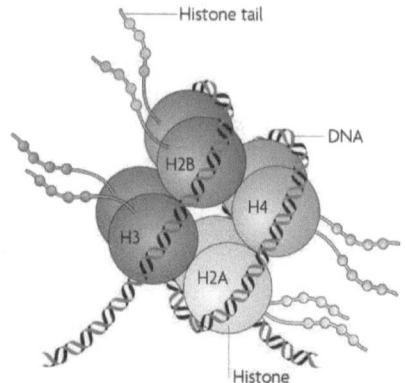

Fig. 2 Schematic representation of the nucleosome structure
The components of a single nucleosome are illustrated. 147-bp of DNA are wrapped around a histone core made of a histone octamere consisting of two copies of the histones H2A, H2B, H3 and H4. Outwards facing side chains represent N-terminal histone tails that can carry different post-translational modifications. The picture is taken from (Tsankova et al., 2007).

The exact mechanism of how these chromatin fibres are packed in higher-order chromosomes to achieve best possible compaction of the up to one giga bp long DNA double strands (Paux et al., 2008) within the nucleus is still unknown. For example, concerning the origin of the 30-nm fibre, two models are discussed, the one-start stack solenoid model and the two-start stack zigzag model (reviewed in (Wu et al., 2007)).

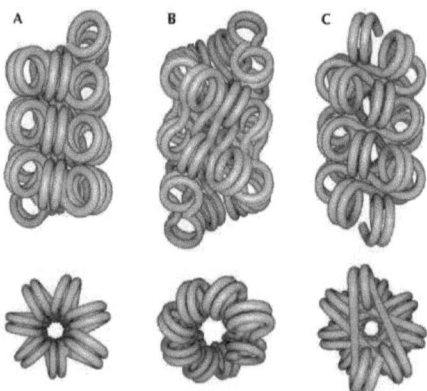

Fig. 3 Illustration of three different models for the arrangement of the 30-nm fibre
Upper illustrations show longitudinal and lower illustrations axial views of the different models. (A) One-start solenoid model. (B-C) Two-start zigzag models as (B) helical ribbon structure or (C) crossed-linker model. The picture is taken from (Wu et al., 2007).

In the solenoid model, the DNA that connects two nucleosomes is bent to a degree that, depending on the length of the linker, six to twelve nucleosomes are required for one helical turn of the chromatin stack to build the 30-nm fibre (Fig. 3A). For the zigzag model, currently two alternative models are proposed. In the helical ribbon model, the linkers are oriented at varying angles along the chromatin fibre (0° to 50°) (Fig. 3B), while the DNA connects nucleosomes across the fibre in the crossed-linker model (Fig. 3C). The length of the linker DNA (~ 10-80 bp) is decisive for the higher order condensation of the unfolded 10-nm chromatin fibre. With optimal, short linkers, the helical ribbon model allows formation of a compact fibre with ~ 10-11 nucleosomes per 11 nm, whereas more heterogeneous or suboptimal linker lengths give rise to the classical 30 nm fibre with approximately six nucleosomes per 11 nm (Fig. 4).

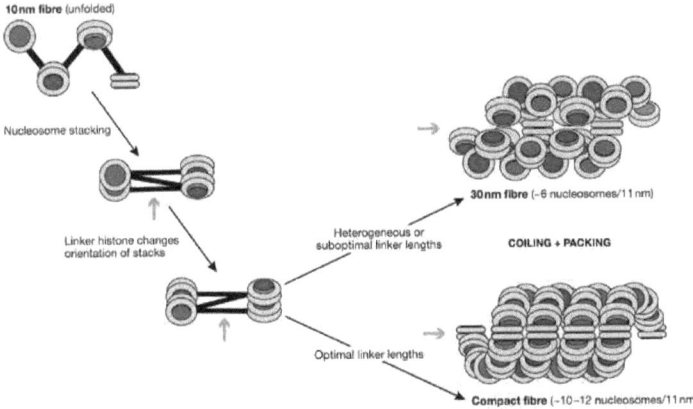

Fig. 4 Models for the formation of the condensed chromatin structure
The different folding mechanisms for a two-start zigzag model are shown that result in varying packaging density. Optimal linker lengths between nucleosomes result in a compact fibre (lower model), while heterogeneous or suboptimal linker lengths lead to the classical 30-nm fibre. Arrows indicate the orientation of the longitudinal axis. The picture is taken from (Wu et al., 2007).

Although a high degree of compaction is necessary for the nuclear organisation of the DNA double helix, the accessibility of the transcriptional machinery must be ensured. To allow this, the eukaryotic genome is organized into regions with open, transcriptionally active euchromatin and regions with condensed, silent heterochromatin. Gene silencing in heterochromatin is not restricted to specific genes, but largely depends on the chromosomal location of a gene, and it involves the establishment of alternative chromatin states that prevent gene expression.

This phenomenon was first discovered in *Drosophila melanogaster* and called position effect variegation (PEV) (Muller, 1930). It was observed that the red eye colour of the flies changed to a red-white mosaic pattern when the *white⁺* gene was translocated from the euchromatic chromosomal position to the vicinity of a heterochromatic region. Another example for gene independent transcriptional silencing of genomic locations is the inactivation of mammalian female X-chromosomes (Avner & Heard, 2001). Here, in each cell, one of the two X-chromosomes is silenced, which is necessary to avoid a double X-chromosomal transcription level in females compared to males that possess only one X-chromosome. Despite events that temporarily affect compacted chromatin states, such as DNA unwinding prior to replication, or DNA repair, silencing is inherited during DNA replication and multiple cell divisions (Ehrenhofer-Murray, 2004). The formation of heterochromatic regions involves extensive, mostly reversible mechanisms, which are regulated by several protein complexes. This control is essential for maintaining or establishing access to the DNA and can be achieved either by enzymes that modify chromatin or by nucleosome remodelling complexes.

1.4 Nucleosome remodelling and histone modifications

The chromatin structure can be changed by three different types of alterations. First, nucleosomes can be altered or repositioned noncovalently by ATP-dependent remodelling complexes (Kingston & Narlikar, 1999; Kornberg & Lorch, 1999; Urnov & Wolffe, 2001; Vignali *et al.*, 2000). Second, histone residues can be modified covalently (Jenuwein & Allis, 2001; Kouzarides, 2002; Wu & Grunstein, 2000). Third, core histones can be replaced by histone variants such as CENP-A (H3) (Smith, 2002) or H2A.X / H2A.Z (H2A) (Redon *et al.*, 2002).

1.4.1 Remodelling complexes

As mentioned before, most of the DNA double helix is tightly wrapped around histone octameres, such that only the outward-facing sequences as well as the DNA linkers are directly accessible to transcription factors. In order to allow DNA-binding proteins to interact with binding sites, which are blocked due to the chromatin structure, processes facilitating conformational changes are required. Chromatin remodelers can shift nucleosomal positions and thereby generate variable nucleosome arrangements in the course of time (reviewed in (Becker & Horz, 2002)). This allows transcription factors to gain access to promoters that might have been occupied by nucleosomes before. Several different remodelling complexes are known, with some families conserved from yeast to human (reviewed in (Sif, 2004)). Next to a catalytic ATPase subunit, all remodelling complexes contain a helicase motif for DNA unwinding activity as well as varying co-factors. At least five families of chromatin remodelers are conserved in all eukaryotes (SWI/SNF, ISWI, NURD/Mi-2/CHD, INO80 and SWR1). The first identified chromatin remodelling complexes belong to the yeast SWI/SNF family (Winston & Carlson, 1992), which, together with the ISWI family, is the best-studied group of remodelers. Chromatin remodelling families show unique protein composition and differ in function. For example, while members of the ISWI family play a role in chromatin assembly following DNA replication and equal spacing of nucleosomes, SWI/SNF remodelers are involved in the reorganisation of nucleosomes to enable transcription factor binding and disordering of even chromatin structures. To date, a wide range of biological functions is attributed to nucelosome remodelling complexes. For example, the *S. cerevisiae* RSC complex, which is a member of the SWI/SNF family, amongst other processes, has been shown to be involved in polymerase II and polymerase III regulation, cell signalling, spindle-assembly checkpoint, chromosome segregation, sister chromatid cohesion, double strand break repair, and cell-cycle progression (reviewed in ((Saha *et al.*, 2006)). As chromatin remodelling has been extensively analysed, several mechanisms have been identified that allow transcription factors to access nucleosomal DNA. Depending on the remodelling complex, different processes can change the DNA-histone conformation, thus enabling DNA accessibility (reviewed in (Jiang & Pugh, 2009)). For example, the SWI/SNF complex creates DNA loops on the histone surface, which leads to an exposure of transcription factor binding sites contained in these sequences. Regulatory sites that are located at the nucleosomal borders can become accessible by the influence of Isw2-containing remodelling complexes that provoke nucleosome sliding. Here, nucleosomal movement loosens the contact between the DNA and histones and exposes the DNA. Furthermore, transcriptions factors can also gain access to the DNA after nucleosome eviction by the RSC complex and histone chapeones such as Asf1.

1.4.2 Histone modifications

As it is the case for remodelling complexes, different kinds of posttranslational modifications of N-terminal histone tails can be distinguished. These modifications comprise phosphorylation of serine residues, methylation of arginine and lysine residues, lysine acetylation and ubiquitination as well as ADP-ribosylation and sumoylation (Berger, 2002;

Freiman & Tjian, 2003; Iizuka & Smith, 2003; Strahl & Allis, 2000). These modifications influence diverse biological processes such as the regulation of gene expression by providing binding sites for transcription factors and silencing by heterochromatin formation (reviewed in (Berger, 2002)), replication (Vogelauer *et al.*, 2002), DNA damage repair (reviewed in (Dinant *et al.*, 2008)) and apoptosis (Ahn *et al.*, 2005). The properties of each modification are determined by the so-called histone code (Jenuwein & Allis, 2001). Most prominent are lysine acetylations by histone acetyltransferases (HATs) and methylations by histone methyltransferases (HMTs), which are reversible by histone deacetylases (HDACs) and histone demethylases (HDMs). While lysines can only be acetylated or deacetylated at the ε-amino group, the residues can be tri-, di-, mono- or non-methylated. Depending on the location of the lysine residue, methylations can serve as activating (H3K4, H3K36, H3K79) or repressing (H3K9, H3K27, H4K20) marks (reviewed in (Kouzarides, 2007)).

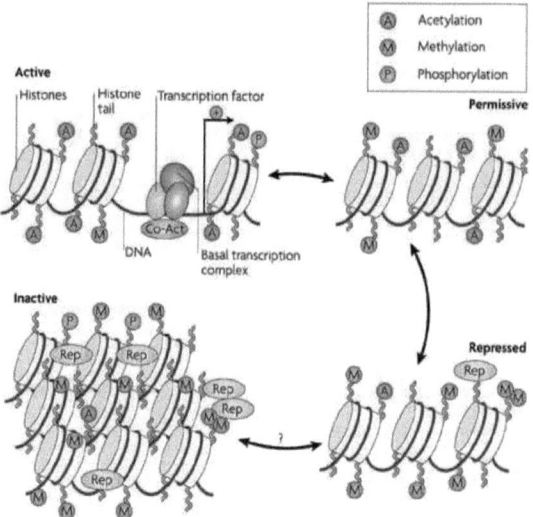

Fig. 5 Illustration of the conversion of chromatin states by N-terminal modifications
Transcriptionally active euchromatin (upper left corner) is associated with acetylation of N-terminal histone tails. The resulting open chromartin conformation allows binding of transcription factors and gene expression. Inactive heterochromatin (lower left corner) is enriched in histone methylations and lacks acetylation. The condensed conformation leads to gene silencing. Changes in the histone modifications and binding of repressors (Rep) or co-activators (Co-Act) can cause a switch from one structural state to the other, as indicated by arrows. The picture is taken from (Tsankova *et al.*, 2007).

Acetylated lysines are usually associated with transcriptionally active euchromatin, while histone deacetylation is a mark for inactive heterochromatin (Fig. 5). At the N-terminal histone H4 tail, four lysine residues (K5, K8, K12 and K16) can be acetylated, which are thought to neutralize its basic charge and therefore reduce DNA-binding and alter the interaction of histones with regulatory proteins or other histones (Grant & Berger, 1999; Roth, 1995). Interestingly, the sites are not equally acetylated. In *S. cerevisiae*, 12 % of the histone H4 tails are not acetylated at all, 36 % are only acetylated at one lysine (preferentially K16 with $^4/_5$ of all cases), 28 % at two and 12 % at all four N-terminal lysine residues (Smith *et al.*, 2003). Unlike in the histone H3 tail, where different sites can be the target for opposing marks like H3K4me / H3K9ac (activating) or H3K4ac / H3K9me (inactivating) (reviewed in (Kim & Workman, 2010)), the *S. cerevisiae* histone H4 N-terminal serine, arginine and lysine residues are only known to be capable of one specific post-translational modification (Fig. 6).

Fig. 6 Representation of known N-terminal histone H4 modification in *S. cerevisiae*
Residues of the N-terminal tail of histone H4 are given in the single amino acid code. Acetylation, methylation and phosphorylation sites are indicated by coloured symbols and the positions from the N-terminal end are numbered.

Detailed studies on histone modification revealed that there is a broad range in target-specificity of the histone-modifying enzymes. On the one hand, many of these enzymes are recruited to special promoters, for example the HDAC Hst1 via Sum1 (Robert *et al.*, 2004) or the HAT Esa1 via general transcription factors like Rap1 or Abf1 (Reid *et al.*, 2000). On the other hand, HDACs (like Rpd3) and HATs can also globally regulate transcription independently of sequence-specific transcription factors (Kurdistani *et al.*, 2002). Furthermore, other enzymes, like the HDAC Sir2, modify well-defined chromosomal regions (reviewed in (Ekwall, 2005)). This indicates a wide variety of regulatory processes controlled by chemical modifications of histone tails. With this in mind, it is not surprising that there is also an interdependence of post-translational histone modifications and nucleosome remodelling. In some cases, a nucleosomal shift might be necessary to allow histone-modifying enzymes to gain access to the previously blocked modification site, while in other cases, prior to gene expression, a remodelling complex might act as the trigger for an activating histone modification (reviewed in (Felsenfeld & Groudine, 2003)).

1.5 Heterochromatic regions in *S. cerevisiae*

Nucleosome remodelling and chromatin modifications are essential regulatory processes for gene activation and silencing. Studies of silencing in the yeast *Saccharomyces cerevisiae* have been fundamental in understanding the mechanisms of gene repression. In *S. cerevisiae*, there are three silenced regions: 1) the two silent mating-type loci *HML* and *HMR*, 2) the telomeres and 3) the ribosomal DNA (rDNA locus) (reviewed in (Rusche *et al.*, 2003)). Formation of heterochromatin in these regions follows a basic pattern that is conserved from yeast to higher eukaryotes. In general, due to limited accessibility, the nucleosomes are less acetylated and DNA replication takes place late in S-phase. Notably, in higher eukaryotes, there is also a reduced activity of DNA-modifying enzymes in heterochromatic regions. A structural element of heterochromatin is the so-called SIR (silent information regulator) complex with the NAD$^+$-dependent histone deacetylase Sir2 as the key component.

1.5.1 Silencing at the silent mating type loci *(HM)*

Haploid *S. cerevisiae* cells assume either **a** or α mating-type, as determined by alternative alleles of the mating type (*MAT*) locus, which is located centromere-proximal on chromosome III. Only haploid cells of opposite mating-types are able to mate and form diploids, which allows recombination of the genetic material of two parental strains. The *MAT***a** mating type is determined by expression of the *MAT***a**1 gene, while expression of the *MAT*α1 and *MAT*α2 genes gives rise to α cells (Herskowitz *et al.*, 1977). These genes encode factors that are responsible for the functional differences of the two mating types, which represent a simple evolutionary form of the more complex sexual differentiation present in higher organisms.

Additional copies of the mating-type genes are found at the so-called silent mating-type loci *HML* and *HMR* located on the left and right arm of chromosome III, respectively. They carry α (*HML*) and **a** (*HMR*) mating-type information that, in contrast to the mating-type information at *MAT*, is permanently repressed. Silencing is mediated by regulatory sequences known as silencers (Loo & Rine, 1995). Both *HM* loci are flanked by an E- (essential) and an I- (important) silencer that differ in sequence, but contain common silencer elements (Fig. 7). While the E-silencer alone can cause silencing of *HML* and *HMR* in the absence of the I-silencer, the I-silencer is only sufficient for *HML*, but not for *HMR* silencing (Abraham *et al.*, 1983; Mahoney & Broach, 1989).

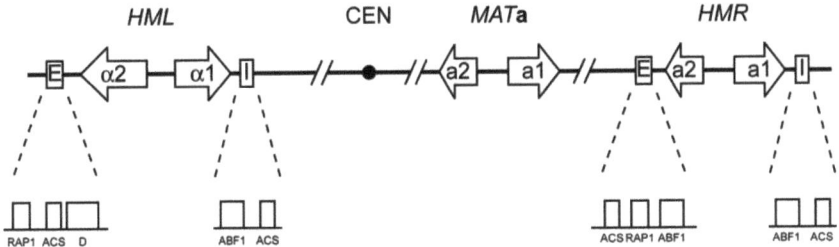

Fig. 7 Schematic representation of the *S. cerevisiae* HM loci
Regions on chromosome III coding for mating-type information are illustrated. The black circle marks the centromere, white arrows the **a**1, **a**2, α1 and α2 mating-type genes located at the mating-type locus *MAT* or the silent mating-type loci *HML* (α) and *HMR* (**a**). White boxes represent the essential (E) and important (I) *HM* silencers flanking the *HM* loci. Binding sites for Rap1, ORC (ACS), Sum1 (D) and Abf1 within the silencer elements are enlarged at the bottom of the illustration. Diagonal gaps represent sequences between the four shown regions.

HM silencing requires multiple *cis*-acting elements within the silencers that are binding sites for DNA-binding proteins and serve as recruitment sites for heterochromatic proteins (reviewed in (Rusche *et al.*, 2003)). Notably, all four silencers contain an ARS consensus sequence (ACS), which is a binding site for the origin recognition complex (ORC) (Foss *et al.*, 1993; Micklem *et al.*, 1993). The I-silencers both contain an additional Abf1 binding site, and the *HMR*-E silencer contains an Abf1 and a Rap1 binding site in addition to the ACS (reviewed in (Loo & Rine, 1995)). *HML*-E consists of three functional elements, a Rap1 binding site, the ACS and a 93-bp sequence, the D element, which are required for silencing (Mahoney *et al.*, 1991). A recent molecular analysis of the D element narrowed it down to a 10-bp core element, termed D2, which is bound by Sum1 (Irlbacher *et al.*, 2005).
In order to establish *HM* silencing, Orc1 recruits the silent information regulator Sir1 to the silencers (Gardner *et al.*, 1999). This leads to the recruitment of Sir4 via its interactions with Rap1 and Sir1, and finally to binding of Sir2 and Sir3 (Rusche *et al.*, 2002). The NAD^+-dependent histone deacetylase Sir2 removes acetyl groups from the N-terminal histone tails of nearby nucleosomes (Imai *et al.*, 2000) and thus provides new binding sites for the Sir2/Sir3/Sir4 (SIR) complex, which requires deacetylated histones in order to bind to chromatin (Hecht *et al.*, 1995). This process results in a positive feedback loop, which leads to the formation of heterochromatin across the *HM* loci (Hoppe *et al.*, 2002; Rusche *et al.*, 2002) (Fig. 8).

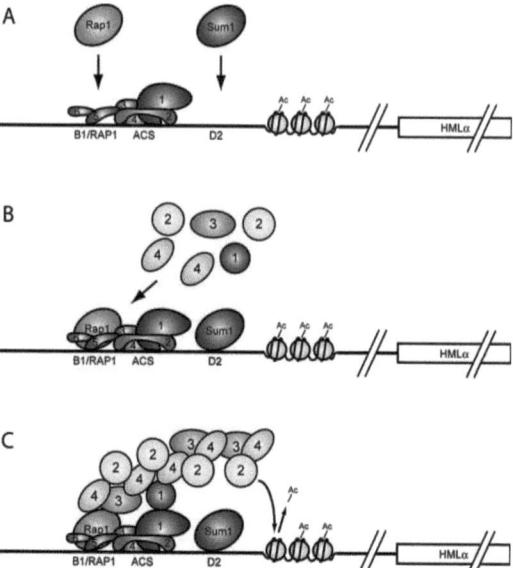

Fig. 8 Model for heterochromatin formation at *HML*
The course of putative events during *HML* silencing is shown. (A) The six-subunit ORC complex assembles on the DNA. Orc1 and Orc4 bind to the ACS, Orc2 and Orc3 between the ACS and the B1 element and Orc5 proximal to B1 (Rap1 binding site). After ORC, the auxiliary factors Rap1 and Sum1 bind to the B1 and D2 elements. (B) Orc1 interacts with Sir1, and Rap1 binds Sir3 and Sir4, which in turn recruit the histone deacetylase Sir2. (C) SIR interactions nucleate the complex and Sir2 deaceylates nucleosomes, which enables new Sir3/Sir4 binding and spreading of the complex leading to *HML* silencing. (ORC binding is modified after (Bell & Dutta, 2002).)

The spreading of silent chromatin into euchromatic regions is hindered by chromatin boundaries (reviewed in (Oki & Kamakaka, 2002)). For instance, the histone acetyltransferase complex SAS-I serves as such a boundary factor in that it antagonizes Sir2 by acetylating H4 K16 (Kimura *et al.*, 2002; Suka *et al.*, 2002). Among others, histone methylation also restricts heterochromatin spreading. H3 K79 methylation by Dot1 (Ng *et al.*, 2002; van Leeuwen *et al.*, 2002) inhibits SIR binding on the nucleosome and thus may prevent the propagation of SIR complexes along the chromatin fiber (Martino *et al.*, 2009).

The *HM* silencers exhibit considerable functional redundancy, because the deletion of any single element, for instance in *HMR*-E (Brand *et al.*, 1985) or *HML*-E (Mahoney *et al.*, 1991), has no measurable effect on repression. Only the simultaneous mutation of two elements, or the combination of the inactivation *in cis* of the binding site with a mutation *in trans* in the gene encoding a second binding factor, causes strong derepression (Sussel *et al.*, 1993). Furthermore, there are several close matches to the ACS around the *HM* loci (Loo & Rine, 1995), and cryptic origins of replication have been described close to *HMR*-E that become activated when the ACS of *HMR*-E is mutated, but do not display silencing activity (so-called non-silencer replicators, (Palacios DeBeer & Fox, 1999)). Only by the removal of this redundancy has it been possible to genetically identify some of the silencer binding factors, for instance ORC (Foss *et al.*, 1993) and Abf1 (Loo *et al.*, 1995b).

Interestingly, all four known silencer binding factors have functions outside of silencing. ORC functions as the eukaryotic replication initiator and is required for initiation at chromosomal origins throughout the genome (reviewed in (Bell & Dutta, 2002)). Rap1 binds to telomeres and functions in telomeric silencing and telomere length regulation (Conrad *et al.*, 1990; Lustig *et al.*, 1990). It also binds to many gene promoters and serves as transcriptional activator (Planta *et al.*, 1995). Abf1 binds some replication origins, and it also contributes to transcriptional activation by binding to gene promoters (Planta *et al.*, 1995). Sum1 is part of a histone deacetylase complex that controls the expression of meiotic genes (Xie *et al.*, 1999) as well as replication initiation of a number of chromosomal origins (Irlbacher *et al.*, 2005; Weber *et al.*, 2008). Several origins are known to exhibit a dual role in silencing and replication initiation (reviewed in (Rehman & Yankulov, 2009)). However, *HMR*-E (ARS317) but not *HML*-E (ARS301) functions as a chromosomal origin of replication. *HML*-E is capable of serving as a replication origin on plasmids, and the chromosomal *HML* locus is replicated by another origin in the vicinity (Vujcic *et al.*, 1999).

1.5.2 Telomeric and rDNA silencing

Telomeres constitute the ends of all eukaryotic chromosomes and are characterized by a unique nucleotide composition. They consist of short, single stranded DNA overhangs and a region of tandem C_{1-3}/TG_{1-3} repeats, which contain several Rap1 binding sites and are defined by the absence of nucleosomes (Gilson *et al.*, 1993; Wright *et al.*, 1992). Due to their structural nature, telomeres would be target to degradation, end-fusions with other telomeres or homologous recombination. However, this is prevented by the formation of heterochromatin at the telomeric ends of the chromosomes. Like at the silent mating-type loci, Rap1 plays an important role in heterochromatin formation, but unlike at the *HM* loci, 10-20 Rap1 proteins serve as a recruitment factor for the SIR complex. First, a Sir2/Sir4 heterodimer is bound followed by Sir3. The complex then spreads as described for *HM* silencing by deacetylating nucleosomes flanking the nucleosome free region, which causes further spreading up to 3 kb into subtelomeric regions (Luo *et al.*, 2002). Reporter genes that are inserted near a region consisting of telomeric C_{1-3}/TG_{1-3} repeats can also be silenced elsewhere in the genome, although less efficiently (Stavenhagen & Zakian, 1994). This can be explained by the presence of additional binding sites in auxiliary elements at natural telomeres. The repeat sequence element CoreX, which is found at all telomeres, contains, depending on the chromosome, multiple Abf1 binding sites as well as an ACS. This can stabilize silencing via Orc1 binding by forming a telomeric loop (Strahl-Bolsinger *et al.*, 1997) and enhance the discontinuous telomeric heterochromatin formation in the vicinity of this element (Pryde & Louis, 1999).

In contrast to *HM* and telomeric silencing, heterochromatin formation at the rDNA locus does not require Sir1 and Sir3 (Smith & Boeke, 1997). The 9.1 kb region encoding the *S. cerevisiae* ribosomal RNA consists of 100-200 repeats (Petes & Botstein, 1977). Sir2 silences the majority of the 35S rDNA, which serves as a precursor for the 25S, 18S and 5.8S ribosomal RNA, at a given time (Smith & Boeke, 1997). Unlike at the other heterochromatic regions, Sir2 here is part of the nucleolar RENT (regulator of nucleolar silencing and telophase) complex (Straight *et al.*, 1999).

1.6 Replication in *Saccharomyces cerevisiae*

Genome duplication by DNA replication is fundamental for the propagation of genetic material in all organisms. Eukaryotic chromosomes are replicated from multiple start sites called replication origins that initiate bidirectional DNA replication.

1.6.1 Replication Origins

Replication initiation at these origins is best understood in the yeast *Saccharomyces cerevisiae*, where approximately 400 origins are used to replicate the DNA of the 16 chromosomes (reviewed in (Bell & Dutta, 2002)). For comparison, about ten thousand origins are thought to exist in the human genome (Gilbert, 2001). The ability of yeast origins to provide initiation and thus autonomous replication to plasmids has allowed the functional dissection of origin elements by measuring plasmid maintenance rates and has coined the term autonomous replicative sequence (ARS).

Plasmid maintenance studies have revealed that yeast origins have a modular structure. They all share a so-called ARS consensus sequence (ACS: WTTTAYRTTTW), which is a binding site for the origin recognition complex (ORC), the replication initiator. However, there is no common DNA sequence known to be present in all eukaryotes. Therefore, *S. cerevisiae* remains the best model organism to analyse DNA replication. The six-subunit ORC complex binds to the origins in an ATP-dependent manner and, together with Cdc6 and Cdt1, recruits the MCM complex, which is likely to be the replicative helicase, to form the pre-initiation complex (reviewed in (Bell & Dutta, 2002)). However, an ORC binding site alone is not sufficient to generate an origin. The ARS1 origin additionally contains three B elements that are required for full initiation (Marahrens & Stillman, 1992). The sequence closest to the ORC binding site, B1, cooperates in ORC binding and DNA unwinding (Lee & Bell, 1997), and B2 is required for loading of the MCM complex (Wilmes & Bell, 2002; Zou & Stillman, 2000). Interestingly, the B3 site is a binding site for the protein Abf1, which functions as a transcription factor elsewhere in the genome (Diffley & Stillman, 1988). The precise function of Abf1 in initiation is not known, but may include a role in nucleosome positioning and origin site selection (Lipford & Bell, 2001). The involvement of transcription factors in initiation seems to be more general, because other transcription factors, Rap1 and Mcm1, have also been identified as origin binding factors that bind in the vicinity of ORC (C site, (Fig. 9)) and influence initiation (Chang *et al.*, 2004; Kimmerly *et al.*, 1988). Also, tethering acidic activators to origins improves initiation (Li *et al.*, 1998), suggesting that transcription factors have a general role in replication initiation.

Notably, individual ARS elements within the yeast genome share very little sequence conservation outside of the ACS. This observation supports the notion that yeast replication origins, in addition to ORC, bind several different auxiliary factors, among them transcription factors that aid in replication initiation, thus explaining why consensus sequences cannot easily be recognized. To determine origins of replication, several large-scale experiments have been performed. For instance, the simultaneous binding of the ORC complex and the replicative helicase MCM serves as an indication for origin function, since both complexes are necessary for replication initiation (Wyrick *et al.*, 2001). Furthermore, replication timing experiments helped to identify origins of replication. Those genomic regions that are replicated earlier in S-phase than adjacent sequences were presumed as origins (Raghuraman *et al.*, 2001). However, all these experimental data are still not sufficient to precisely map all *in vivo* origins, as the influence of other regulatory factors is crucial for initiation of replication. The following fact illustrates this discrepancy between potential and true *in vivo* origins. While the *S. cerevisiae* genome comprises approximately 12,000 ARS consensus sequences, with approximately 400 only 3.3 % of these sequences serve as replication origins (Nieduszynski *et al.*, 2006). The remaining 97 % do not initiate DNA replication despite the fact that they contain an ACS. In the current model, different subsets of origins are bound by different replication modulators that support full initiation of these origins. One interesting goal thus is to identify additional DNA binding factors that might serve as auxiliary factors for replication initiation.

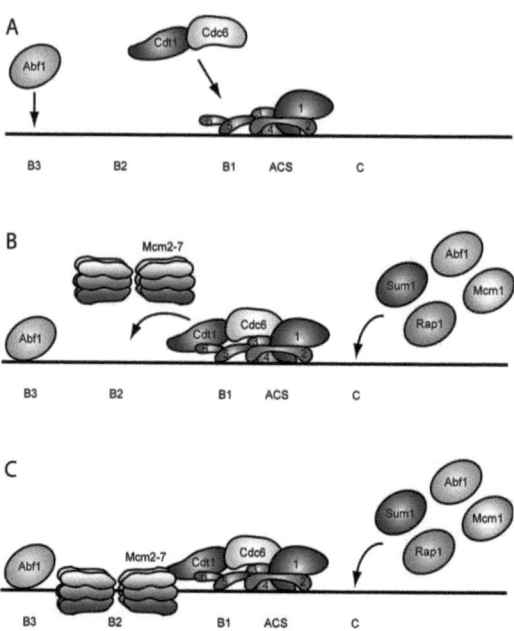

Fig. 9 Schematic representation of the process of replication initiation
Protein binding to the DNA prior to replication initiation is illustrated. (A) After binding to the ACS and B1 element, the ORC complex recruits the auxiliary factors Cdt1 and Cdc6, and Abf1 binds to the B3 element. (B) To form the pre-replication complex, the double-hexameric MCM2-7 replicative helicase complex is recruited. (C) The MCM complex binds to the B2 element for DNA-unwinding and further transcription and auxiliary factors such as Sum1 and Rap1 bind to the C element.

1.6.2 Replication origins in other eukaryotes

Less is known about replication origins in other organisms. In contrast to *S. cerevisiae* there is no ACS in fission yeasts, but an enrichment of A-T nucleotides is found in ARS sequences (Clyne & Kelly, 1995; Okuno *et al.*, 1999). This common feature is missing in metazoans. Here, initiation takes place at large initiation zones (Dijkwel *et al.*, 2002) as well as in short defined regions (Toledo *et al.*, 1998). A possible explanation for the need for distinct consensus sequences is that efficient DNA replication best takes place at intergenic regions where less conflict potential with DNA transcription is probable. As the *S. cerevisiae* genome is highly transcribed, a random arrangement of origins would statistically affect many areas, which are also bound by the transcriptional machinery. Therefore, it would be evolutionarily beneficial to target replication origins to these intergenic sequences (Brewer, 1994). Organisms with more non-transcribed regions such as metazoans are less forced to initiate replication at certain chromosomal locations and therefore do not require well-defined consensus sequences. Interestingly, it has been found in metazoans that efficient, solitary origins are located near clustered genes (Abdurashidova *et al.*, 2000; Aladjem *et al.*, 1998; Toledo *et al.*, 1998), while multiple inefficient origins are found at large, non-transcribed regions (Dijkwel *et al.*, 2002; Ina *et al.*, 2001; Little *et al.*, 1993).

1.6.3 Regulation of initiation function

The lack of consensus sequences, with the exception of *S. cerevisiae*, for binding of the pre-replicative complex (pre-RC) allows only speculation about how exactly origin function is determined in other eukaryotes. Two solutions to this problem seem possible. First, chromosome structure or epigenetic factors such as chromatin organization and modifications may determine the assembly of the pre-RC and define active origins (reviewed in (DePamphilis *et al.*, 2006)). Notably, it has been demonstrated that several ORC subunits can be modified in a cell cycle-specific manner, thus altering chromatin affinity and stability in metazoan cells (DePamphilis, 2005). In *S. cerevisiae*, several cases are known where time of origin firing was dependent on chromosomal localization. Late initiating origins such as ARS501 have been shown to fire early when placed on a plasmid (Ferguson *et al.*, 1991) as well as some early origins to fire late when located proximal to telomeres (Raghuraman *et al.*, 2001). Second, a selection of only few replication origins out of a far bigger amount of potential origins is supported by the observation that approximately 10-fold more Mcm2-7 complexes assemble on DNA than ORC molecules or active origins exist (Blow & Dutta, 2005). Orc1 seems to be the key player that determines the sites for pre-RC assembly when cells enter G_1-phase, while the other ORC subunits remain bound to chromatin throughout the cell cycle (reviewed in (DePamphilis *et al.*, 2006)). The abundance of many potential replication origins can be explained by the fact that it is crucial that on the one hand the whole genome is replicated during cell cycle and on the other hand replication of every sequence is limited to one event per cell division. Therefore, about 90 % of origins licensed in G_1 remain inefficient or inactive in S-phase. However, origin efficiency can adapt to changes in chromatin structure and association with the nuclear matrix in G_1 (Courbet *et al.*, 2008). In

S. pombe as well as in other eukaryotes, so-called replication factories have been found, which contain several replication forks and are subject to changes in number and subnuclear distribution during S-phase (reviewed in (Costa & Blow, 2007)). Replication origins that are normally replicated passively by neighbouring origins can serve as a back-up mechanism when the progression of the replication fork is inhibited for example due to DNA damage sites and initiate replication (reviewed in (Blow & Ge, 2008)).

1.7 The correlation between silencing and replication initiation

Although heterochromatin formation at the silent mating type loci *HML* and *HMR* by silencing and DNA-replication are distinct cellular processes, both share some common features. Assembly of the six-subunit ORC complex at the ACS plays an essential role as well as the binding of the auxiliary factors such as Rap1 or Sum1 (compare Fig. 8 and Fig. 9).

1.7.1 Dual role of ORC

So far, all tested *S. cerevisiae* silencers have a dual nature, which allows them to act in silencing and replication initiation, although pre-RC at silencers fire inefficiently (reviewed in (Rehman & Yankulov, 2009)). In the context of silencing, the function of ORC is assumed to follow a distinct pattern (reviewed in (Weinreich *et al.*, 2004)), and the exact mechanism of what determines whether ORC binding to the ACS causes replication initiation or silencing is still unknown. For instance, it is speculated that the B1 element might be responsible for ORC function, although several experiments provided contradictory results in that mutations in the B1 element, depending on the tested ARS element, sometimes seemed to impair silencing but not replication and vice versa (reviewed in (Rehman & Yankulov, 2009)). Additionally, a temporal separation of ARS function is discussed as well as cell cycle-dependent phosphorylation of ORC and MCM subunits. While phosphorylation of the MCM proteins during S-phase initiates origin firing (Blow & Dutta, 2005; Stillman, 2005), ORC phosphorylation by the cyclin Clb5 can prevent reestablishment of the pre-RC (Nguyen *et al.*, 2001; Vas *et al.*, 2001; Wilmes *et al.*, 2004) and might facilitate interaction of ORC with the SIR complex. In this context, recruitment of Clb5 by Orc6 would switch ORC function from replication to silencing.

1.7.2 Sum1 in *HM* silencing and replication initiation

In previous work, the Sum1 protein was identified as a novel auxiliary initiation factor (Irlbacher *et al.*, 2005). In contrast to the transcription factors like Rap1 or Abf1 described above, Sum1 in other contexts functions as a transcriptional repressor. It binds upstream of a number of middle sporulation genes and represses them during vegetative growth by recruiting the histone deacetylase (HDAC) Hst1 to the promoter, thus providing chromatin-mediated gene repression (Pierce *et al.*, 2003; Xie *et al.*, 1999). In this function, Sum1 is part of a protein complex containing Hst1 and the bridging repression factor of MSEs, Rfm1 (McCord *et al.*, 2003).

In addition to its repressor function, Sum1 shows several links to ORC-mediated replication initiation as well as repression of the silent mating-type loci *HML* and *HMR*. The deletion of *SUM1 (sum1Δ)* is synthetically lethal with a conditional mutation in *ORC2, orc2-1*, which causes an initiation defect (Irlbacher *et al.*, 2005; Suter *et al.*, 2004). This suggests that a number of origins require Sum1 as an auxiliary factor, such that cells cannot tolerate the loss of Sum1 when ORC function is compromised. Accordingly, a number of Sum1-dependent origins have been identified (Irlbacher *et al.*, 2005; Lynch *et al.*, 2005). Sum1 also shows a weak physical interaction with ORC (Irlbacher *et al.*, 2005). Interestingly, a mutant version of Sum1, Sum1-1, was identified that bestows upon Sum1 an improved ability to interact with ORC (Rusche & Rine, 2001; Sutton *et al.*, 2001). Sum1-1 thus is aberrantly recruited to a number of origins, among them the silent mating-type locus *HMR*, where it establishes Hst1-dependent gene silencing (Lynch *et al.*, 2005; Rusche & Rine, 2001; Sutton *et al.*, 2001). Natural Sum1 binds to the *HML*-E silencer and, in cooperation with other silencer-binding factors, promotes gene silencing at *HML* (Irlbacher *et al.*, 2005).

1.7.3 Influence of histone modifications on origin activity

Like all metabolic processes on DNA, replication initiation in eukaryotic cells must contend with the packaging of the DNA into chromatin, which generally restricts access to the DNA. Origin function has been shown to depend on the chromosomal context and the positioning of nucleosomes by ORC. Nucleosomes proximal to ORC facilitate the initiation of replication, whereas covering of the origin by nucleosomes interferes with initiation (Lipford & Bell, 2001). Furthermore, the SWI/SNF chromatin remodelling complex in some contexts is required for full stability of plasmids with a minimal origin, and tethering of an activator to an origin can create dependence of the origin on a chromatin remodeler (Flanagan & Peterson, 1999).

Replication initiation is also influenced by histone acetylation. It changes timing of origin firing in that the absence of the HDAC Rpd3 causes late origins to fire early, whereas tethering the histone acetyltransferase Gcn5 to late origins advances their time of firing (Aparicio et al., 2004; Vogelauer et al., 2002). This suggests that the deacetylated chromatin state suppresses early initiation. Furthermore, histone acetylation affects the efficiency of replication initiation at a subset of origins. The absence of the HDAC Sir2 partially suppresses the initiation defect of a cdc6 mutation, indicating that initiation at some origins is more efficient when the chromatin is in the acetylated state (Crampton et al., 2008; Pappas et al., 2004).

The Sum1/Rfm1/Hst1 deacetylase complex

Three different classes of histone deacetylaes in *S. cerevisiae* are known (reviewed in (Ekwall, 2005)). Class I comprises Rpd3, Hos1 and Hos2, class II Hda1 and Hos3 and the Sirtuin class III contains besides Sir2 the homologues of Sir two (HSTs). The HST family contains four different histone deacetylases (Hst1-4). With 63 % overall and 82 % conserved core identity, Hst1 shows the strongest homology to Sir2. Hst1 is found in two separate complexes, together with the histone deacetylase Hos2 in the SET3 complex and as the sole HDAC in the Sum1/Rfm1/Hst1 complex (Pijnappel et al., 2001; Xie et al., 1999). In this context, Sum1 binds in the vicinity of ORC and recruits the histone deacetylase Hst1 via the bridging factor Rfm1 (Fig. 10). Despite of the strong homology, Sir2 and Hst1 have at least some different functions. In other cases, Hst1 can partially compensate for a disruption of *SIR2*. This was shown by experiments with *hst1Δ* cells, which (unlike *sir2Δ* cells) did not display defects in *HMLα* silencing (Brachmann et al., 1995) or a rDNA recombination phenotype (Derbyshire et al., 1996). However, Hst1 overexpression could partially rescue *HMRa* silencing defects in a *sir2Δ* strain (Brachmann et al., 1995).

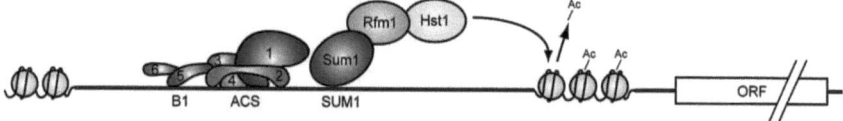

Fig. 10 Model for histone deacetylation by the Sum1/Rfm1/Hst1 complex
The schematic representation shows Sum1-mediated histone deacetylation by Hst1. The ORC complex binds to the ARS consensus sequence (ACS) and the B1 element. Sum1 binds in the vicinity of ORC followed by recruitment of the histone deacetylase Hst1 via the bridging factor Rfm1. N-terminal histone tails of adjacent nucleosomes are then deacetylated by Hst1.

1.8 Outline of this thesis

Although the initiation of DNA replication and gene silencing by heterochromatin formation are two distinct processes, both share key components. In this study, the regulation of origin function and silencing of the silent mating type locus *HML* in *Saccharomyces cerevisiae* were analysed.

The aim of this thesis was to investigate the contribution of ORC and Sum1 to the regulation of silencing of the yeast silent mating type locus *HML* and replication initiation of origins that are simultaneously bound by ORC and Sum1. It was determined whether the hitherto characterized silencer elements of *HML*-E were necessary and sufficient for *HML* silencing. To this end, a synthetic version of the *HML*-E silencer that consisted of the Rap1 and ORC binding sites and the D2 element, the binding site for Sum1 (Irlbacher *et al.*, 2005), alone was designed and characterized. This silencer provided strong repressing activity, showing that these elements are sufficient for *HML* silencing. Furthermore, it was sensitized towards mutations *in cis* as well as *in trans*. Additionally, the synthetic *HML*-E silencer was used to identify novel factors that contribute to *HML* silencing. With Dot1 and Sir1, two regulators were found, since *dot1Δ* and *sir1Δ* caused *HML* derepression in this sensitized background.

In a second part of this thesis, it was asked how Sum1, which is part of a complex with Rfm1 and the NAD^+-dependent histone deacetylase Hst1, exerted its function in replication initiation. It was found that both *rfm1Δ* and *hst1Δ* were synthetically lethal with *orc2-1*, showing that the Sum1/Rfm1/Hst1 HDAC complex was required for Sum1's initiation function. Seven ARS elements were identified whose initiation capacity depended on Sum1 and Hst1. In their absence, acetylation at lysine 5 of histone H4 was significantly increased at these origins. Also, mutation of the acetylatable lysines in the H4 tail to imitate the acetylated state caused reduced initiation of these plasmids. Taken together, these results show that Sum1 recruited the HDAC Hst1 to selected origins in the yeast genome, and that histone deacetylation by Hst1 at these origins was required for their full initiation function.

Besides Sum1, other factors (among these Dat1, Gat3 and Rgm1) have been proposed in bioinformatics analyses (T. Manke, personal communication) to bind to the same intergenic regions as ORC and therefore might also contribute to origin activity. Interestingly, deletions of any of those factors were synthetically lethal together with a mutation in the ORC complex, *orc2-1*. However, the cause for this lethality remained unclear, since neither replication initiation of the respective origins nor silencing function was affected by those factors.

2 Methods

2.1 *E. coli* strain

DH5α   F⁻ φ80d*lac*ZΔM15 Δ(*lac*ZYA-*arg*F)U169 *rec*A1 *end*A1 *hsd*R17(rk⁻, mk⁺) *pho*A *sup*E44 *thi*-1 *gyr*A96 *rel*A1 λ⁻ (Invitrogen, electro competent).

2.2 *E. coli* growth conditions

E.coli strains used for plasmid amplification were cultured according to standard procedures (Sambrook *et al.*, 1989) at 37 °C in Luria-Bertani (LB) medium (5 g/l yeast extract, 10 g/l tryptone, 5 g/l NaCl) supplemented with 100 µg/ml ampicillin.

2.3 *Saccharomyces cerevisiae* media and growth conditions

Media were as described previously (Sherman, 1991). Yeast minimal medium (YM: 6.7 g/l yeast nitrogen base w/o amino acids or YM(msg): 1.7 g/l yeast nitrogen base w/o amino acids and w/o ammonium sulfate with 1 g/l L-glutamic acid monosodium salt) was supplemented with 2 % glucose and as applicable with 20 µg/ml for adenine, histidine, methionine, tryptophan and uracil or 30 µg/ml leucine and lysine. YM + 5-FOA (5-fluoro-orotic acid) *URA3*-counterselective medium contained 1 mg/ml 5-FOA. YPD-G418 plates contained 200 mg/l geneticin and were used to select for *KanMX* knock-outs. Selection for haploid *MATa* cells in the screen with the *S. cerevisiae* deletion library was performed with YM or YM(msg) plates containing 50 mg/l canavanine and 50 mg/l thialysine. Unless otherwise indicated strains were grown in liquid full medium (YPD: 10 g/l yeast extract, 20 g/l peptone, 2 g/l glucose) or on YPD plates (20 g/l agar) at 30 °C.

2.4 *S. cerevisiae* strain construction

The yeast strains used in this study were generated by transformation with plasmids (Klebe *et al.*, 1983), direct gene disruption, chromosomal integration or crossing of haploid *MATa* and *MATα* cells or derived from the laboratory strain collection and are listed in Table 1.

2.4.1 Gene disruption

Gene disruptions were performed using the *KanMX* or *HisMX* cassette according to the guidelines of EUROFAN (Wach *et al.*, 1994) or by replacing the complete open reading frame with a PCR-generated *URA3MX* sequence.

2.4.2 Chromosomal integrations

Construction of *S. cerevisiae* strains with synthetic *HML*-E alleles
A two-step replacement strategy was used for the generation of the *URA3MX*-marked synthetic *HML*-E alleles. First, an *URA3MX* cassette was chromosomally integrated at the *HML*α locus, replacing an approximately 4-kb region flanking wt *HML*-E, thus resulting in *hml*Δ::*URA3MX* strains. Second, plasmids carrying synthetic *HML*-E (see 2.5.2) were digested with the restriction endonucleases *Apa*LI and *Hin*dIII to create 3.8-kb fragments containing the mutated *HML*-E locus and adjacent regions. Via homologous recombination, the constructs replaced the introduced *URA3MX* in the *hml*Δ strain (AEY3387) to create *ura3*⁻ strains with synthetic *HML*-E alleles, which were truncated by 144 bp of flanking regions compared to the wt *HML*-E sequence. Integrants were identified by counterselection on 5-FOA medium and verified by PCR analysis. In order to mark the synthetic *HML*-E allele for genetic crosses, an *URA3MX* cassette was introduced into the *HML-SS* ΔI strain 5' of the gene *VBA3* in the direct vicinity of *HML*-E (Chr. III coordinates 8966-9065) via a third homologous recombination resulting in $URA3^+$ cells. All gene disruptions and sequence insertions were verified by PCR analysis.

Insertion of telomeric silencing constructs
For the insertion of synthetic telomeric constructs, TEL VII L::*URA3* (pJR1543) and TEL VII L::*ADE2-URA3* (pJR1544) *Eco*RI/*Sal*I or *Not*I/*Sal*I digested linear fragments were chromosomally integrated into *S. cerevisiae* strains by homologous recombination.

2.4.3 Crossing, sporuation and tetrad dissection of *S. cerevisiae* strains

Parental strains of opposing mating types were mixed in a drop of YPD medium and grown on YPD over night at 30 °C. Cells were streaked on YM plates supplemented with amino acids selecting for diploids and grown for two to three days. Diploids were grown for ten hours on YPD and subsequently incubated for three days at 30 °C or five days at 23 °C for sporulation induction on sporulation medium (Sherman, 1991) or for the deletion library screen: 10 g/l KAc, 1g/l yeast extract, 0.05 % glucose, 20 g/l agar supplemented with 7.5 µg/ml leucine, 5 µg/ml histidine and methionine). For asci digestion a loop of sporulated cells was incubated for 6.5 min in 10 µl zymolyase buffer (1 M sorbitol, 0.1 M NaCitrate, 60 mM EDTA, pH 8.0, 5 mg/ml zymolyase). Zymolyase digestion was stopped by adding 100 µl H₂O. Tetrads were dissected on YPD plates using a micromanipulator (Narishige) connected to a Zeiss Axioscope FS microscope. The ascospores were incubated for two to three days at 30 °C or 23 °C. Marker analysis was achieved by replicating the cells with the help of sterile velvet cloth on different selection plates.

Table 1: *Saccharomyces cerevisiae* strains used in this study

Strain [a]	Genotype
AEY2	MATa can1-100 ade2-1 his3-11,15 leu2-3,112 trp1-1 ura3-1 (=W303-1A)
AEY3	MATα can1-100 his3-11,15 leu2-3,112 trp1-1 ura3-1 (W303-1B, but lys2Δ ADE2)
AEY10	MATα sir1::LEU2 ADE2
AEY24	MATa ade2-101 his3-11,15 trp1-1 leu2-3,112 ura3-1 can1-100 orc2-1 rho⁰
AEY232	MATα HMR-SS ΔI ade2-101 his3-11,15 trp1-1 leu2-3,112 ura3-1 can1-100 orc2-1 +pLEU2
AEY264	MATα his4
AEY265	MATa his4
AEY760	MATα ΔAhmr::TRP1 rap1-12(LEU2) ade2-101 his3-11,15 trp1-1 leu2-3,112 ura3-1 can1-100
AEY1341 [1]	MATa smc3-42
AEY1499	AEY2 hst1Δ::KanMX
AEY1880	MATa hht1-hhf1Δ::LEU2 hht2-hhf2Δ::HIS3 ade2 lys2Δ::His4 his3-11,15 trp1-1 leu2-3,112 ura3-1 can1-100
AEY2189	MATα hir1Δ::KanMX cac1Δ::LEU2 LYS
AEY2579 [*]	MATα hht1-hhf1Δ::LEU2 hht2-hhf2Δ::HIS3 ade2 lys2Δ::His4 his3-11,15 trp1-1 leu2-3,112 ura3-1 can1-100 hmr::ADE2
AEY2900	MATα asf1Δ::KanMX LYS
AEY3054 [*]	MATα can1Δ::MFA1pr-HIS3-MFα1pr-LEU2 his3Δ leu2Δ0 ura3Δ0 met15Δ0 lys2Δ0
AEY3358	AEY2 sum1Δ::HisMX
AEY3387	AEY2 hmlΔ::URA3
AEY3388	AEY2 HML-ΔI
AEY3398	AEY2 HML-E ACS⁻ ΔI
AEY3595	AEY1 TEL VII L ::ADE2::URA3
AEY3705	MATα dot1Δ::KanMX sas2Δ::TRP1
AEY3915	AEY3 dat1Δ::KanMX
AEY3917	AEY3 gat3Δ::KanMX
AEY3928	AEY3 rfm1Δ::KanMX
AEY3940	MATa rfm1Δ::KanMX orc2-1 +pRS316-ORC2 LYS
AEY3941	MATα rfm1Δ::KanMX orc2-1 +pRS316-ORC2 LYS ADE2
AEY3947	AEY3940 + pRS315
AEY3948	AEY3941 + pRS315
AEY3957	AEY3940, but pLEU2-ORC2
AEY3958	AEY3941, but pLEU2-ORC2
AEY3968	MATα gat3Δ::KanMX orc2-1 +pRS316-ORC2
AEY3970	MATα gat3Δ::KanMX orc2-1 +pRS316-ORC2 LYS
AEY3971	MATa dat1Δ::KanMX orc2-1 +pRS316-ORC2 ADE2
AEY3972	MATα dat1Δ::KanMX orc2-1 +pRS316-ORC2 ADE2
AEY3973	AEY2, but hht1-hhf1Δ::LEU2 hht2-hhf2Δ::HIS3, lys2Δ::hisG + pCEN4-TRP1 HHF1-HHT1
AEY3974	AEY2, but hht1-hhf1Δ::LEU2 hht2-hhf2Δ::HIS3, lys2Δ::hisG + pCEN4-TRP1 hhf1-10(H4K5,8,12,16Q) HHT1
AEY3975	AEY3968, but pRS315-ORC2
AEY3976	AEY3968 + pLEU2

AEY3977	AEY3970, but pRS315-*ORC2*
AEY3978	AEY3970 + pRS315
AEY3979	AEY3971, but pRS315-*ORC2*
AEY3980	AEY3971 + pRS315
AEY3981	AEY3972, but pRS315-*ORC2*
AEY3982	AEY3972 + pRS315
AEY4022	AEY3 *rgm1Δ::KanMX*
AEY4023	AEY3 TEL VII L *::URA3*
AEY4024	AEY3915 TEL VII L *::URA3*
AEY4025	AEY3917 TEL VII L *::URA3*
AEY4026	AEY3 TEL VII L *::ADE2::URA3*
AEY4027	AEY3915 TEL VII L *::ADE2::URA3*
AEY4028	AEY3917 TEL VII L *::ADE2::URA3*
AEY4033	*MATa dat1Δ::KanMX LYS*
AEY4043	*MATa gat3Δ::KanMX LYS*
AEY4077	*MATa rgm1Δ::KanMX orc2-1* +p RS316-*ORC2* (W303-1A)
AEY4078	*MATa rgm1Δ::KanMX orc2-1* +p RS316-*ORC2 LYS* (W303-1A)
AEY4079	AEY4077, but pRS315-*ORC2*
AEY4080	AEY4077 + pRS315
AEY4081	AEY4078, but p*LEU2-ORC2*
AEY4082	AEY4078 + pRS315
AEY4087	AEY1341 + pRS316-*SMC3*
AEY4088	AEY1341 + pRS316-*SMC3*
AEY4120	AEY2 TEL VII L *::URA3*
AEY4121	AEY4033 TEL VII L *::URA3*
AEY4122	AEY4043 TEL VII L *::URA3*
AEY4123	AEY2 TEL VII L *::ADE2::URA3*
AEY4124	AEY4033 TEL VII L *::ADE2::URA3*
AEY4125	AEY4043 TEL VII L *::ADE2::URA3*
AEY4128	AEY4087 *sum1Δ::HisMX*
AEY4129	AEY4088 *sum1Δ::HisMX*
AEY4140	*MATa hst1Δ::KanMX orc2-1* + pRS315-*ORC2* + p RS316-*ORC2* (W303-1A)
AEY4142	*MATa hst1Δ::KanMX orc2-1* + pRS315 + pRS316-*ORC2* (W303-1A)
AEY4156	AEY4160 *dat1Δ::KanMX*
AEY4157	AEY4160 *gat3Δ::KanMX*
AEY4158	AEY4161 *dat1Δ::KanMX*
AEY4159	AEY4161 *gat3Δ::KanMX*
AEY4160 [2]	*FEP100-5 URA3@IX L site1*
AEY4161 [2]	*FEP100-10 URA3@IX L site1*
AEY4199	AEY3 *ctf18Δ::KanMX*
AEY4202	*MATa hht1-hhf1Δ::LEU2 hht2-hhf2Δ::HIS3 hmr::ADE2 orc2-1* +p*URA3-HHT1-HHF1* (light red)
AEY4203	*MATa hht1-hhf1Δ::LEU2 hht2-hhf2Δ::HIS3 hmr::ADE2 orc2-1* +p*URA3-HHT1-HHF1* (red)
AEY4206	*MATa ctf18Δ::KanMX sum1Δ::HisMX*
AEY4207	*MATα ctf18Δ::KanMX sum1Δ::HisMX ADE2*
AEY4208	*MATa ctf18Δ::KanMX sum1Δ::HisMX ADE2 LYS*
AEY4239	AEY4202 + p*CEN4-TRP1 HHF1-HHT1* (wt)
AEY4241	AEY4202 + p*CEN4-TRP1 hhf1-10(H4K5,8,12,16Q) HHT1*
AEY4402 [*]	AEY3054 *hmlΔ::URA3*

AEY4404	AEY3387 + synth. wt *HML*-E ΔI (*Apa*LI/*Hin*dIII fragment from pAE1396)
AEY4406	AEY3387 + *HML-SS* ΔI (*Apa*LI/*Hin*dIII fragment from pAE1386)
AEY4408	AEY3387 + *HML-SS* rap⁻ ΔI (*Apa*LI/*Hin*dIII fragment from pAE1390)
AEY4410	AEY3387 + *HML-SS* acs⁻ ΔI (*Apa*LI/*Hin*dIII fragment from pAE1388)
AEY4412	AEY3387 + *HML-SS* D2⁻ ΔI (*Apa*LI/*Hin*dIII fragment from pAE1392)
AEY4416	AEY4402 + synth. wt *HML*-E ΔI (*Apa*LI/*Hin*dIII fragment from pAE1396)
AEY4417[a]	AEY4402 + *HML-SS* ΔI (*Apa*LI/*Hin*dIII fragment from pAE1386)
AEY4427	AEY4404 *sum1Δ::HisMX*
AEY4428	AEY4406 *sum1Δ::HisMX*
AEY4462	AEY4404 *rap1-12::LEU2*
AEY4464	AEY4406 *rap1-12::LEU2*
AEY4466	AEY4404 *orc2-1*
AEY4469[a]	AEY4417 5'VBA3/*HML URA3MX*
AEY4486	*MATa HML*-E ss ΔI *orc2-1* (W303)
AEY4537	*MATa sir1::LEU2 ADE2* (W303)
AEY4538	*MATa sir1::LEU2 HML-SS* ΔI *LYS* (W303)
AEY4570	*MATa HML-SS* ΔI 5'VBA3/*HML URA3MX ADE2 TRP1*
AEY4573 [3]	*MATα can1Δ::STE2pr-his5 lyp1Δ ura3Δ leu2Δ his3Δ met15Δ0*
AEY4629	AEY4573 5'VBA3/*HML URA3MX*
AEY4752	*MATa HML*-E ss ΔI 5'VBA3/*HML URA3MX LYS*
AEY4754[a]	*MATa hir2Δ::KanMX HML-SS* ΔI 5'VBA3/*HML URA3MX*
AEY4805	AEY4752 *dot1Δ::KanMX*
AEY4807	AEY4752 *hir2Δ::KanMX LYS*

[a] Unless indicated otherwise(*), strains were isogenic to W303 and constructed during the course of this study or originate from the laboratory collection
Indices give the source of strains: [1] C Michaelis, [2] E. Louis, [3] F. van Leeuwen.

Table 2: Oligonucleotides used for knock-outs and molecular cloning

Oligonucleotide	DNA sequence 5'-3' [a]
CTF18.S1	CCT AAT GTG TAC ACT ATT TGA CCC AAA AGG TGG ATG TAA GGT CAG GGA TC C GTA CGC TGC AGG TCG AC
CTF18.S2	CAA GTA TGC TTC TTA AGA GAG ACT GCG TAT ATA TCT TAC GTC ATT TAT TCA TCG ATG AAT TCG AGC TCG
CTF18-ko-contr_fw	GCA ACA ACA TAC TGT AAC CAT TGT G
CTF18-ko-contr_rv	CTA TGC GCT ATA TTG TAT CCC ACA G
Cyc8.S1	GAC TAG TAC TAC AAC TAC AAC AGC AAC AAC AAC AAA CAA AAC ACG ACT GGG ACG TAC GCT GCA GGT CGA C
Cyc8.S2	GTT GAA CAG ATT GTA GTT GTT GCT GTT GAG GTG GCA GTT GAG AGC TTT GCG CAT CGA TGA ATT CGA GCT CG
Cyc8-ko_contr_fw	CCG ATT ATC AAA GCA AAA GCG C
Cyc8-ko_contr_rv	CGT TAT TCC TGT TGC GGG AGC TTG
DAT1.S1	GAT CAC CTT GTG AAT CTA CAA ACT HTC CTA AAG TAT ATT GGA GCA GGA CAT GGG GCG TAC GCT GCA GGT CGA C
DAT1.S2	GTG GCA TAT ACG AAT GTT TTA GTG GTA TGC TGG AAA TGA ATG TCA TAT GGT TGC GAT CGA TGA ATT CGA GCT CG
DAT1.up	GGC TTC TAC CAG TCG CGT TTC AAG C
DAT1.down	GCA GGT TCT TCA GGA CCT GAA TG

GAT3.S1	<u>CAT TAA ATG TAA ACA ATT ATC AAC TAG AAG CAA ATA TAA AGC CAG AAG GAA GAA T</u>CG TAC GCT GCA GGT CGA C
GAT3.S2	<u>GCT TTG ACA TAA GTA TAT AAC ATT CCG AGC AGA AAT AAA TTC TCT TAA CGC GTT C</u>AT CGA TGA ATT CGA GCT CG
GAT3.up	ACA TTA CCT GCT TAG CCG CCT GCC
GAT3.down	CAG TCC ATT GAG AAG TAT GCC
his5 S.p. up	CTA TGG GTA ACT TTG CCG G
HIR2.S1	<u>CCA TAC AGA GGA ATA CGC CAC GCA GCA AAG GAG TTC CTA CAC AAT CCG GAA CAG A</u>CG TAC GCT GCA GGT CGA C
HIR2.S2	<u>GAA TAT AAT GAA AAA TAT AAG AGT TTA AAC TAT ACA TTG TTA AAG CCA AAC TAA GGC</u> ATC GAT GAA TTC GAG CTC G
Hir2-ko_contr_fw	CCA CGT ATC TGG AAT GAC CAA C
Hir2-ko_contr_rv	CCT GTG TTG GCA TTG GTA TGC
HML-E down II	GTT TAC ATT TCA TTC TAT GTG CGC TAG
HML-E Rekomb-kontr.up	GAG TCT CAT TTG GGG AAG ATT CG
HML-For-control	GGT ACA ACT CTT GGT GGT GTG C
HML-for-contr-II	CTT GGA TTT GCT CTA CAA GCA TCG C
HML-KO-1	<u>CCG CCT CCT TTC ACA ACA AAG TAT CAC GAG CTC ATC TAG AGC CTT ACG AAG G</u>GA TTG TACTGA GAG TGC AC
HML-KO-2	<u>GCT TAC TTC ATT TAT TAGA TAA TAT AAG GTA CAG TGT TCA TGA ATT TTT CTC ATG TTG</u> CTG TGC GGT ATT TCA CAC CG
HML-rev-control	GAG ATC GAA AGA AAG CTC CCG C
HST1.S1	<u>CGA ACA CTT CTC TTC TTT TTT GTT GTT TTT GTG AGA AAA AAA AAT CTA A</u>CG TAC GCT GCA GGT GCG AC
HST1.S2	<u>CTC CCC CTT CTG TGT TTT CTT CTT TTT TTT TTT TTT TTT TTT GGA AT</u>A TCG ATG AAT TCG AGC TCG
HST1-ko-contr_fw	GGT GAA CGC CAC TCA GTT GGC C
HST1-ko-contr_rv	CGC AAG ATA CTA AAG AAG AGT CGC GC
KanMX-K2	GCC CCT GAG CTG CGC ACG TC
KanMX-K3	CCC AGA TGC GAA GTT AAG TGC GC
Mrpl20.S1	<u>CTT TCA ACA AGT TAA AAT GGA AAA TTA AAGA CAA AGT AAA ATA GCA CAA GAC</u> GTA CGC TGC AGG TCG AC
Mrpl20.S2	<u>GTA CTT ATA GAG TAG TAT TTA CAC GAT TGT TAT TAT ATT TAT ATA GCA T</u>AT CGA TGA ATT CGA GCT CG
Mrpl20-ko_cont.fw	GGCTTTTGCGCTAGTTTCTGC
Mrpl20-ko_cont.rev	CACCTTCGTCGTGAAAGTCCG
Rfm1-MX-kofw	<u>ATT AAA AGA ATT TAA TTA GAA CAA CAG GAA GGT GTT ATA AGA AAG TGC GAC</u> GTA CGC TGC AGG TCG AC
Rfm1-MX-korv	<u>GAT ACG TCA TAT TTC TCT CTA TTT ATA TTT ATT TAC TTC TTC AAA GAA GC</u>A TCG ATG AAT TCG AGC TCG
RFM1-ko_contr_fw	CCC AAG TTT GCA ATT GGA TCC AAG G
RFM1-ko_contr_rv	GAG TAC GGA TAT TAA GGC ATG CC
RGM1.S1neu	<u>CTA TTC CTC CCC ACT CAC GTC TAT AGT CCA GGA CAA TTA GCA TGA CAC TGG TT</u>C GTA CGC TGC AGG TCG AC
RGM1.S2neu	<u>CAA TAG CAG CAA TGA TGC GGC TTG TGA AAT GGA GGA GTG GGT GCG TA GAG CTT A</u>AT CGA TGA ATT CGA CTC G
RGM1-	GCA CGG ACC ATT AAG TTA CGT G

37

ko_contr_fw	
RGM1-ko_contr_rv	GAG TAA ATA TAA GGG GTC GAG GAA C
Sequenz_ss_HML-E_fw	GGA CGA GAC ACC AGA AGA TAA TTT AG
Sir1 S1	<u>GCT CGG AGC TGG CTA GTG CTG CTT CAT CTT TAC TGT CTC TTA AGG GTC C</u>CGTA CGC TGC AGG TCG AC
Sir1 S2	CAC AGG CCT AGG AAG CTT CTT GAT CTG CCT TGC GAA CAA TGG CAA CGC GAT CGA TGA ATT CGA GCT CG
Sir1-A1-control	GGA TGA GCA GAT CCT TCC GAT
Sir1-A4	CCT CAA ATC CAA TCT AAA TAC AGC
Sir1 ATG.down	CAT CAA TAA CTG CAA GCC TGG AG
ss_HML-E+AflII_up	GCG ctt aag <u>TTT CAA ATT GAT TCA AAC ACC TTT CAC</u>
ss_HML-E(SpeI)down	CCT CGT CAA AAG AAG TCA GGA C
Sum1-kanMX_fw	<u>GAG ATC AAA CGA AAA GTT TCA TAC ATA ATT AAC AAA ATT CGT TTG TTG CGG GGC</u> GTA CGC TGC AGG TCG AC
Sum1-kanMX_rv	<u>CTA TTC TCG AAA CTG CCC CAA CGT ACG GAC CAG CTT AAC GGA TAT CTG GCG GTA TGA</u> TCG ATG AAT TCG AGC TCG
Sum1_kocontr_fw	CTT TCC CAC GTG GCC TTA ACT ACG
Sum1_kocontr_rv	CAA TCC TCG GAG CCA TTC CAG TGC
URA3-ko-contr-rev	GTG CGG TAT TTC ACA CCG CAT AGG G
VBA3/HML.S1	<u>CCG TAA GGG TTA CTG ATA CAC AAT TTC CTT TTT GTA AAG AGT ATT TGA GCC</u> GTA CGC TGC AGG TCG AC
VBA3/HML.S2	<u>CGC GCT GCA TAT GAG GTG CGG CGC TAT CTG TTA AAT ATG TAC CAA TTT GC</u>A TCG ATG AAT TCG AGC TCG
VBA3/HML MX-ko.up	GTG TCT ATA CAC CTG GTG AC

[a] Nucleotides complementary to *S. cerevisiae genomic* sequence, which were used for chromosomal integration by homologous recombintion, are underlined. Lower case letters indicate an introduced *Afl*II restriction endonuclease site.

2.5 Plasmid constructions

All *E. coli* manipulations were carried out according to standard protocols (Sambrook *et al.*, 1989).

2.5.1 ARS fragments

ARS fragments (length approx. 500 bp; ARS446 and ARS447 1500 bp) containing ARS and Sum1 consensus sequences (ACS: WTTTAYRTTTW; SUM1: DSYGWCAYWDW) were amplified via PCR from genomic DNA of wild-type cells and subcloned into the pGEM-T Easy vector (Promega) (for details, see Fig. 27). The vector pAE1076, which contains ARS1012, was used as a backbone for construction of the *CEN4-URA3*-ARS plasmids. It was digested with *Eco*RI and *Hind*III to release the ARS1012 fragment, and the new ARS fragments with compatible overhangs were ligated into the vector. The final constructs were verified by sequence analysis.

2.5.2 *HML*-E fragments

The 3' end of the wild-type *HML* locus (Chr. III Coordinate 11410-12016) containing the 3' flanking region of the E silencer and the 5' sequence of the W element including the *Spe*I restriction site and lacking the I silencer (*HML* ΔI) was amplified from a plasmid (pAE1034) using *Vent*-Polymerase and subcloned. The oligonucleotides were designed to generate an *Afl*II restriction site 5' of this sequence. The 734 bp wild-type *Afl*II/*Spe*I *HML* fragment (Chr. III Coordinate 11182-11915) was then replaced by the truncated 511 bp *Afl*II/*Spe*I sequence (Chr. III Coordinate 11410-11915) using the 5' introduced *Afl*II restriction site of the subcloned PCR fragment to create a truncated *HML* sequence lacking both, the E and I silencers (pAE1351). After restriction endonuclease digestion a 4.8 kb *Bam*HI/*Hind*III fragment this *HML* truncation was brought into a yeast vector (pAE1378) that is based on YCplac22 (CEN, *TRP1*). This vector was used to generate the *HML*-E mutant plasmids via the yeast-based, oligonucleotide-mediated gap repair technique (YOGRT) (DeMarini et al., 2001).

Synthetic single-stranded 79 bp *HML*-E variants (Chr. III, coordinates 11230 to 11302) were designed which carried no mutations, mutations in the sequences flanking and surrounding but not in or in one the following - the Rap1 binding site, the ACS or the D2 element. For each synthetic *HML*-E silencer variant, two additional 55-57 bp oligonucleotides were used that were half complementary to the ends of the synthetic 79 bp sequence and half complementary to ends of the *Afl*II cut pAE1378 vector (sequences given in Table 3). For the yeast-based, oligonucleotide-mediated gap repair technique, 1 µg of all three single-stranded fragments was combined with 100 ng of the *Afl*II digested vector for *S. cerevisiae* transformation (AEY2) to create the synthetic *HML*-E plasmids (pAE1386, pAE1388, pAE1390, pAE1392, pAE1396) through the ability of yeast to undergo homologous recombination. The constructs were verified by sequencing.

Table 3: Oligonucleotides used for generating synthetic *HML*-E fragments

No.	Oligonucleotide	DNA sequence 5'-3' [a]
1	wt_HML-E_fw	tta agA GTA TCT TAT GAA TGG GTT TTT GAT TTT TTT ATG TTT TTT TAA AAC ATT AAA GTT TTC GGC ACG GAC TTA TTT G
2	ss_HML-E_fw	tta ag*T TCG ATA* TAT GAA TGG GTT T*AT TTT GT*T TTT ATG TTT T*AA ATA GAT CTA TAT A*TT TTC GGC ACG GAC *GTT TTA T*
3	ss_HML-E_RAP1mut_fw	tta ag*T TCG ATA* **TTT CAT TCG CTA A***AT TTT GT*T TTT ATG TTT T*AA ATA GAT CTA TAT A*TT TTC GGC ACG GAC *GTT TTA T*
4	ss_HML-E_ACSmut_fw	tta ag*T TCG ATA* TAT GAA TGG GTT T*AT TTT GT***A TAA GGC GCC G***AA ATA GAT CTA TAT A*TT TTC GGC ACG GAC *GTT TTA T*
5	ss_HML-E_D2mut_fw	tta ag*T TCG ATA* TAT GAA TGG GTT T*AT TTT GT*T TTT ATG TTT T*AA ATA GAT CTA TAT* **AAA TAC CGG AGG CAG** *GTT TTA T*
6	YOGRT wt HML-E.up	AAT CAA AAA CCC ATT CAT AAG ATA CTc tta agA AAT TAC ATT CCA TTG CGA TAC ACC
7	YOGRT wt HML-E.down	GGT GTT TGA ATC AAT TTG AAA CTT AAC AAA TAA GTC CGT GCC GAA AAC TTT AAT G
8	YOGRT ssHML-E.up	*CAA AAT AAA CCC ATT CAT ATA TCG AA*c tta agA AAT TAC ATT CCA TTG CGA TAC AC

9	YOGRT ssHML-E.down	GGT GTT TGA ATC AAT TTG AAA CTT AA*A TAA AAC* GTC CGT GCC GAA AAT *ATA TAG*
10	YOGRT RAP1mut.up	*CAA AAT **TTA GCG AAT GAA ATA** TCG AA*c tta agA AAT TAC ATT CCA TTG CGA TAC ACC
11	YOGRT D2mut.down	GGT GTT TGA ATC AAT TTG AAA CTT AA*A TAA AAC **CTG CCT CCG GTAT** T*TA TAT *AG*

[a] Sequences complementary to the synthetic *HML*-E fragments are underlined. Nucleotide mutations compared to wild-type sequence affecting binding sites for Rap1, the ORC complex and Sum1 are indicated by bold italic letters, mutations of surrounding sequences are printed in italics. Lower case letters indicate parts of or complete *Afl*II restriction sites.

Table 4: Combination of oligonucleotides used to generate synthetic *HML*-E fragments

synthetic *HML*-E silencer	Oligonucleotide No. from Table 3
wt *HML*-E (79 bp) ΔI	1, 6, 7
HML-SS ΔI	2, 8, 9
HML-SS rap⁻ ΔI	3, 9, 10
HML-SS acs⁻ ΔI	4, 8, 9
HML-SS D2⁻ ΔI	5, 8, 11

Table 5: Plasmids used in this study

Plasmid	Description	Source [a]
pAE179	*mat*Δ*::URA3*	
pAE478	pFA6a-*KanMX*	
pAE568	pRS426-*SMC3*	
pAE929	pFA6a-*HIS3MX*	
pAE1034	pUC18 *HML*α *HML*-ΔI	
pAE1076	*CEN4-URA3* + ARS1012	O. Aparicio
pAE1081	*CEN4-URA3* + ARS1013-3	O. Aparicio
pAE1126	*CEN4-URA3* + ARS606	
pAE1130	*CEN4-URA3* + ARS1223	
pAE1135	*CEN4-URA3* + ARS1511	
pAE1192	pRS314-*HHT1 hhf1-10* = *H4 K5, 8, 12, 16 Q*	R. Morse
pAE1193	pRS314-*HHF1-HHT1*	
pAE1240	*CEN4-URA3* + ARS433	
pAE1242	*CEN4-URA3* + ARS607	
pAE1243	*CEN4-URA3* + ARS1109	
pAE1250	*CEN4-URA3* + ARS446	
pAE1252	*CEN4-URA3* + ARS447	
pAE1276	*CEN4-URA3* + ARS1001	O. Aparicio
pAE1277	*CEN4-URA3* + ARS1002	O. Aparicio
pAE1278	*CEN4-URA3* + ARS1009	O. Aparicio
pAE1279	*CEN4-URA3* + ARS1014	O. Aparicio
pAE1280	*CEN4-URA3* + ARS1025	O. Aparicio
pAE1315	pRS315-*ORC2*	
pAE1316	pRS316-*ORC2*	
pAE1351	pAE1034 cut *Afl*II/*Spe*I + (Chr. III Coordinate 11410 to 11920)	
pAE1378	YCplac22 cut BamHI/HindIII + *HML*-E ΔI from pAE1351	

pAE1386	(Chr. III Coordinate 11410 to 14562) pAE1378 + *HML-SS* (Chr. III, Coordinates 11230 to 11302 mutated)	
pAE1388	pAE1378 + *HML-SS* acs⁻ (Chr. III, Coordinates 11230 to 11302 mutated)	
pAE1390	pAE1378 + *HML-SS* rap⁻ (Chr. III, Coordinates 11230 to 11302 mutated)	
pAE1392	pAE1378 + *HML-SS* D2⁻ (Chr. III, Coordinates 11230 to 11302 mutated)	
pAE1396	pAE1378 + wt *HML*-E (Chr. III, Coordinates 11230 to 11302 mutated)	
pAE1484	YCplac33 + 4,8 kb BamHI/HindIII cut pAE1386 *HML-SS* ΔI	
pAG60	*URA3MX*	
pJR1543	TEL VII L::*URA3*	D. Gottschling
pJR1544	TEL VII L::*ADE2-URA3*	D. Gottschling
pRS315	*CEN6-LEU2* + ARSH4	(Sikorski & Hieter, 1989)
pRS316	*CEN6-URA3* + ARSH4	(Sikorski & Hieter, 1989)
pRS406	*URA3*	
YCplac22	*CEN4-TRP1*	
YCplac33	*CEN4-URA3*	

[a] Unless indicated otherwise, plasmids were from the laboratory collection or constructed during the course of this study.

2.6 Plasmid maintenance assay

Plasmid loss rates were determined for wild-type (AEY2), *sum1Δ::HisMX* (AEY3358) and *hst1Δ::KanMX* (AEY1499) or wild-type (AEY3), *dat1Δ::KanMX* (AEY3915) and *gat3Δ::KanMX* (AEY3917) carrying *CEN-URA3* plasmids containing the different ARS elements as follows. Yeast transformants were grown to stationary phase in liquid minimal medium lacking uracil, and cultures were used to inoculate YPD supplemented with adenine, histidine, leucine, lysine, tryptophan and uracil (or histidine, leucine, lysine and tryptophan). Cells were grown for at least 12 doublings at 30 °C with shaking. Before and after the incubation, equal amounts of the cultures were plated on minimal medium with or without uracil. The plasmid loss rate (L) was determined by measuring the fraction of cells containing the plasmid before (F_i) and after (F_f) incubation in full medium as $1-10^x$ with $x = [\log(F_i) - \log(F_f)]$ / number of doubling times (McNally & Rine, 1991). The loss rate is therefore equivalent to the fraction of daughter cells that have received no plasmid during the previous cell division.

Table 6: Oligonucleotides used for the construction of ARS plasmids used in plasmid maintenance assays

Oligonucleotide	DNA sequence 5'-3'
ARS433_pLoss_fw	GAA TTC CTA TTG TTG CTT TAG TTT CTG TGC AC
ARS433_pLoss_RV	AAG CTT GCA CAC AAT GCC AAA TGC TCC
ARS446-pLoss_neu_fw	GAA TTC GCT CAT GCA GGT ATT TCA AAC C
ARS446-pLoss_neu_rv	AAG CTT CGA CAG CAA AGG CAG AG
ARS447-pLoss_neu_fw	GAA TTC CAG GAA GAC CAT TGA TCG AAG G
ARS447-pLoss_neu_rv	AAG CTT CGT CGA GGA CAA AGT AAA CCT G
ARS606fw	GTC TTC TTG ATA ATT CTG TGG GCG C
ARS606RV	GTC TTG CCT TAG GAC TCA GCC AGG
ARS607-pLoss_fw	GAA TTC GAG TCA GGT CGA TCC TGC TAT TG
ARS607-pLoss_rv	AAG CTT CTT GGT AAT CAA GGC TAG AAG TGT AC
ARS1109-pLoss_fw	GAA TTC CAG TAC CCT CTT GAT GTT CT TGC
ARS1109-pLoss_rv	AAG CTT GCA GAA GAC ATT ATC TGC CAT GC
ARS1223up	CTT GAG TCA AGT TCA GAG TAA TTT TCG G
ARS1223down	CCC ATT TGA CGC AAG GCA ATT CCC CTG
ARS1511up	CGA CCC TGC AGC AGC TGC TCA G
ARS1511down	CCA GCT CAT CTG CAG CTG CC
ARSH4-pLoss_fw	GAA TTC GAG ACA AGG TAG AAC CTT ATA CGG
ARSH4-pLoss_rv	AAG CTT CCG AAT TGT TTC ATC TTG TCT GTG

2.7 Antibodies

The following antibodies were used in this study (rabbit antiserum, upstate): Anti-acetyl-Histone H4 (Lys5 Catalog # 07-327 Lot # 30417); Anti-acetyl-Histone H4 (Lys12 Catalog # 07-595 Lot # 28885); Anti-acetyl-Histone H4 (Lys16 Catalog # 07-329 Lot # 32214); Anti-acetyl-Histone H4 (polyclonal antiserum Catalog # 06-866 Lot # 20667).

2.8 Chromatin immunoprecipitation (ChIP)

100 ml of yeast cells were grown to an OD_{600} of 1 and cross-linked with 1 % formaldehyde for 30 minutes with shaking at room temperature. Cells were harvested by centrifugation (three minutes, 4,000 x g) and washed twice in 1 x TBS. Pelleted cells were resuspended in 100 µl ice-cold lysis buffer (50 mM HEPES (pH7.5), 1 mM EDTA, 140 mM NaCl, 1 % (v/v) Triton X-100 and 0.1 % sodium deoxycholate) containing protease inhibitors and disrupted with glass beads. The cell supernatant was sonicated four times for ten seconds with 200 ms impulses, centrifuged, and the protein concentration was adjusted with lysis buffer. One aliquot was taken as an input control for the quantitative real-time PCR. Aliquots were precleared with protein G-agarose (Sigma) for two hours at 4 °C and incubated over night with 3 µl antibody. After incubation, the lysates were treated with protein G-agarose-beads for five hours at 4 °C. The immunoprecipitates were washed with 1 ml of the following buffers (ice-cold): 1. low salt solution (0.1 % (v/v) SDS, 1 % (v/v) Triton X-100, 2 mM EDTA, 20 mM Tris (pH 8.1) 150 mM NaCl), 2. high salt solution (0.1 % (v/v) SDS, 1 % Triton (v/v) X-100, 2 mM EDTA, 20mM Tris (pH 8.1) 500 mM NaCl), 3. LiCl buffer (0.25 M LiCl, 1 % (v/v) Nonidet P-40, 1 % (w/v) sodium deoxycholate, 1 mM EDTA, 10 mM Tris pH 8.1), twice 1 x TE. The samples and the input DNA were subsequently treated with elution buffer (1 % (v/v) SDS, 0.1 M $NaHCO_3$), incubated over night at 65 °C to reverse

cross-linking and incubated one hour with proteinase K (Roche). The DNA was extracted with chloroform-phenol-isoamylalcohol and precipitated with ethanol.

2.9 Quantitative real-time PCR

The ChIP and the input samples were used in different dilutions as templates for the PCR with SYBR Green RealMasterMix (Eppendorf). The reference dilutions were used to generate a standard curve that was taken to determine the DNA amount of the ChIP samples. Fragments of 211 to 356 bp in size were amplified (see Fig. 27). The real-time PCR setup was as follows: An initial denaturation step at 94 °C for two minutes, followed by 45 cycles of denaturation at 94 °C for 15 seconds, annealing at 56 °C for 30 seconds and elongation at 68 °C for 40 seconds. After cooling to 40 °C for two minutes and one minute at 50 °C, the temperature was raised every five seconds in 1 °C intervals up to 95 °C. The template amount of the immunoprecipitated samples was measured as the mean value of three dilutions relative to the computer-calculated standard curve of the input reference. Evaluation of the data comprised two independent ChIPs with standard error of the mean (six data values).

Table 7: Oligonucleotides used for quantitative real-time PCR of ChIP samples

Oligonucleotide	DNA sequence 5'-3'
ARS433_qPCR.up	CGC ATG TAG ATT TAC CTC TTT TCC C
ARS433_qPCR.down	CAA TAC TTA GCA AAT TGT TCG AGA CG
ARS446_qPCR.up	GGA AAT ATT AAA TGA AGC AGT TGG AAC C
ARS446_qPCR.down	CAA GTT ATA TTT CGG AGC TGT CCC
ARS447_qPCR.up	GAA GAA AGC ATT AGC GTC GTT ACG
ARS447_qPCR.down	GAT GAT ATA ACG TTC AAT TTA ATT GAT GGG C
ARS606_fw	GTC TTC TTG ATA ATT CTG TGG GCG C
ARS606_rv1	GAA ACT CCA GCA GCT TGA GCC AG
ARS607.fw	GGT GAT ATA AAC ACT ACA TTC GC
ARS607.rv	GCT TTC TAG TAC CTA CTG TGC
ARS1109_qPCR.up	GTA TTA ACT TTC AGT AAA GTT ACC CGC C
ARS1109_qPCR.down	GCA ACC TGA AAA TTC ATA GAA CTT TTG G
ARS1223_qPCR.up	TCT GTC TCA TGC ACT TGG AAG C
ARS1223_qPCR.down	CAT TTG ACG CAA GGC AAT TTC CC
ARS1511_qPCR.up	CGT ATT AAT ACA CAA TAA TCT ATC CTC TCA GG
ARS1511_qPCR.down	GAC ATA TTG TGC CTC AAC TCT TGC
ARSH4_qPCR.up	GGG ATT CGT ATT CAA CTG CCC G
ARSH4_qPCR.down	TTC TTC ATT CCG TAA CTC TTC TAC C
CDC20_qPCR.up	GGT AAC CGT TCT GTA CTT TCT ATT GCG
CDC20_qPCR.down	GGA TAT GAA CGA GAA GAG TAT GCC G

2.10 Silencing assays

2.10.1 Telomeric silencing assay

Telomeric silencing was tested using synthetic TEL VII L::*ADE2-URA3* or *URA3* constructs (Gottschling *et al.*, 1990) chromosomally integrated into wild-type (AEY2), *dat1Δ::KanMX*

(AEY4033) and *gat3Δ::KanMX* (AEY4043) strains by homologous recombination. *URA3* silencing was tested by spotting dilutions of the different cells on YM plates with or without uracil and YM + 5-FOA plates and comparing growth on these media.
ADE2 silencing was analysed by observing the colony colour. While repression of the telomeric *ADE2* resulted in red colonies, reduced silencing led from red colonies with white sectors to white colonies with red sectors or a completely white phenotype, depending on derepression intensity. The percentage of these four different colour phenotypes was measured as an indicator for telomeric *ADE2* silencing.

2.10.2 *HML* silencing assay

HML silencing was performed with a *MATa his4* mating-type tester strain (AEY265). For a qualitative mating strains were streaked in 1 cm^2 patches on YPD plates and grown over night before replicating on YM plates with a lawn of 2 OD units of the mating-type tester strain. Cells were incubated for two days at 30 °C. Photographic documentation of the amount of grown diploids allowed to determine the mating efficiency and thereby *HML* silencing. A quantitative mating analysis was performed as described previously (Ehrenhofer-Murray *et al.*, 1997) from three independent experiments.

2.10.3 Deletion library *HML* silencing screen

A yeast strain with the minimal *HML*-E silencer (AEY4629) was used in a genetic screen with the *S. cerevisiae* deletion library to identify factors that influence *HML* silencing in this sensitized background. For this purpose, a previously described synthetic lethal screening procedure (Tong & Boone, 2007) was modified as follows. After selection for haploid *MATa* cells strains were directly pinned on YM(msg)-G418 plates containing canavanine and thialysine and lacking uracil to select for *HML-SS* ΔI *xxxΔ::KanMX* double mutants. After recovery on YPD plates (one day 30 °C) cells were mixed with 100 µl of the *MATα his4* mating-type tester strain (AEY265, OD$_{600}$ ~ 2.5) in microplate wells, pinned on YM plates and grown for two days at 30 °C.

2.10.4 Deletion library transformation with synthetic *HML*-E plasmid

For plasmid transformation in microplates, deletions strains were grown on YPD-G418 plates to inoculated 50 µl YPD+G418 liquid cultures. Cells were incubated with shaking in microplates covered with air-permeable membranes. After two days, 195 µl YPD were inoculated with 5 µl of the cultures and incubated in microplates for 3.5 h with shaking and harvested by brief centrifugation. Transformation with a *CEN4-URA3* + *HML-SS* ΔI plasmid (pAE1484) was performed according to standard protocol (Klebe *et al.*, 1983) with cell amount-adjusted volumes of reagents. Cells were spotted on YM plates containing histidine, leucine and methionine and grown for two days. Single transformants were restreaked on YPD plates and used for a patch mating assay (see 2.10.2). All incubations were performed at 30 °C.

3 Results

3.1 A novel minimal *HML*-E silencer caused *HML* derepression in *sir1*Δ and *dot1*Δ strains

3.1.1 Three *HML*-E core elements were sufficient to establish *HML* silencing

The goal of this study was to determine whether the combination of the three known *HML*-E silencer domains alone was sufficient to establish *HML* silencing. To this end, a synthetic version of *HML*-E was designed that consisted of the Rap1 binding site, the ACS and the D2 element alone and its silencing capacity was tested. In a first step, in order to remove potential binding sites near *HML*-E, a core version of natural *HML*-E was constructed in which 43 bp of upstream (telomere-proximal) and 107 bp of downstream sequence were removed, thus retaining 79 bp of natural *HML*-E (Fig. 11A). The upstream deletion was chosen such as not to disturb the function of the neighbouring *VBA3* gene, and the downstream deletion removed the intervening sequences between the D2 element and the W region of *HML* (Herskowitz *et al.*, 1992). This *HML*-E version, termed wt *HML*-E (79 bp), was introduced into an *HML* allele lacking the I silencer (Irlbacher *et al.*, 2005) in order to measure silencing by the E silencer alone. Silencing of *HML* was determined by measuring the mating ability of *MAT***a** strains carrying the *HML* allele by a patch mating assay (Fig. 12A) as well as by a quantitative mating assay (Fig. 12B). These assays are based on the fact that derepression of *HML* in a *MAT***a** strain causes an **a**/α cell type, thus resulting in the loss of **a**-mating ability of the strain (Herskowitz *et al.*, 1992). In these assays, wt *HML*-E (79 bp) mated as well as a wild-type strain (Fig. 12), indicating that this silencer retained full silencing capacity. This suggested that the sequences flanking this core silencer did not significantly contribute to silencer function.

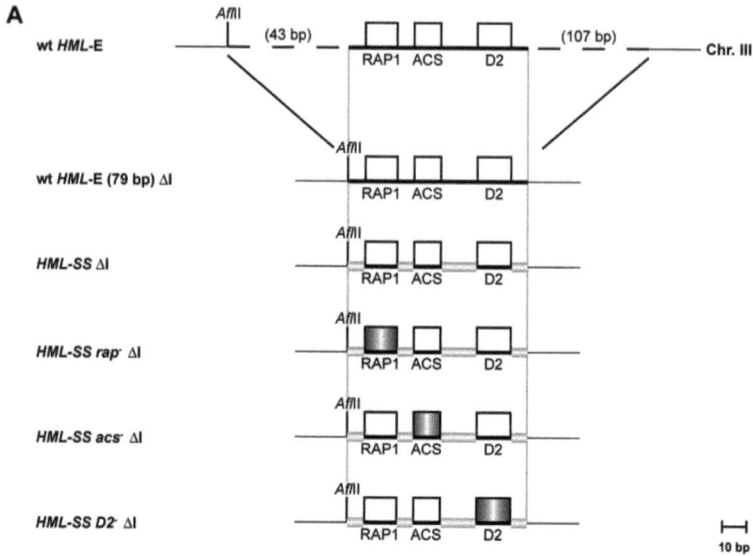

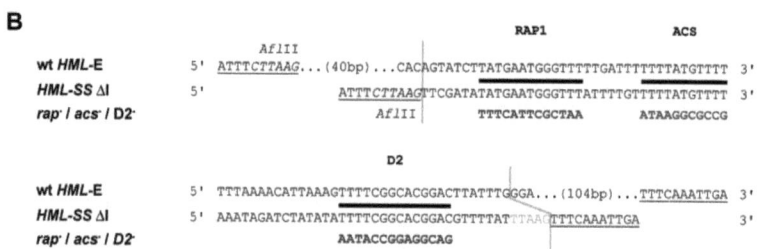

Fig. 11 Design of a synthetic *HML*-E silencer
(A) The wt *HML*-E locus and five truncated constructs are shown. In the synthetic *HML*-E silencer (*HML-SS* ΔI), the nucleotide order between the silencer elements was partially scrambled while retaining the base pair composition (orange boxes). Mutations affecting binding sites for Rap1 (*rap1⁻*, blue), the ORC complex (*acs⁻*, green) and the D2 element (*D2⁻*, red) are indicated by colours. Vertical transparent grey lines delineate the mutated region. (B) DNA sequence of wt *HML*-E (Chr. III, coordinates 11177 to 11420) and synthetic *HML*-E silencer variants. The Rap1, ACS and D2 elements are indicated by black bars. Corresponding mutations in these elements within the *HML-SS* ΔI context are shown in bold letters. Transparent grey lines define the 79-bp synthetic *HML*-E silencer and the corresponding wt sequence. Light grey letters show nucleotides originating from the insertion of an *Afl*II site that are only present in the synthetic constructs. Italics indicate the *Afl*II site.

Next, in order to eliminate potential redundant elements within the 79-bp core *HML*-E silencer, a minimal *HML*-E silencer was generated, which was termed "synthetic silencer" (*HML-SS* ΔI), in which the wild-type Rap1, ACS and D2 elements were retained, whereas the nucleotide order of the short flanking and intervening sequences of the 79-bp *HML*-E truncation was scrambled. The mutations were chosen such that the base composition and the distance between the elements remained unchanged (Fig. 11B). A mating test of a *MAT*a strain with this synthetic *HML-SS* ΔI variant showed a strong mating ability indicative of substantial *HML* silencing. However, it was somewhat reduced as compared to the strain with wt *HML*-E (79 bp) ΔI (Fig. 12A). The quantitative analysis showed that *HML-SS* ΔI retained approximately 60 % silencing ability of wild-type *HML*-E (Fig. 12B). This showed that the combination of binding sites for Rap1 and ORC with the D2 element alone was sufficient to generate strong *HML* silencing. Of note, the silencing provided by synthetic *HML*-E was stronger than that by the synthetic *HMR*-E silencer, which retained approximately 15 % silencing ability compared to wild-type *HMR*-E (Ehrenhofer-Murray *et al.*, 1997). The difference between *HML-SS* and wt *HML*-E (79 bp) indicated that the sequence scrambling had removed unknown functional sequences that contributed 40 % to silencing. In the further experiments, the *HML-SS* ΔI allele was used as the minimal *HML* silencer.

3.1.2 The Rap1 and ORC binding sites and the D2 element were essential for *HML* silencing

The deletion of individual silencer domains in the natural *HML*-E silencer does not cause *HML* derepression (Mahoney *et al.*, 1991), indicating that there is functional redundancy in natural *HML*-E. It was asked whether the synthetic silencer eliminated this redundancy by determining whether the binding sites for Rap1 and ORC as well as the D2 element were required for silencing of *HML-SS* ΔI. To address this, three *HML*-E variants were constructed in which the sequence of one of these three elements was mutated (Fig. 11B). For the *HML-SS rap*⁻ ΔI construct, every other nucleotide of the Rap1 binding site was changed by a transitional mutation. The mutation of the ACS in the *HML-SS acs*⁻ ΔI allele was designed analogous to that in the synthetic *HMR*-E silencer (McNally & Rine, 1991). Furthermore, the mutation of the D2 element in the *HML-SS D2*⁻ ΔI variant was created by transitional mutation of every other nucleotide, as previously described (Irlbacher *et al.*, 2005).

Significantly, all three mutations led to strong *HML* derepression as indicated by a strong loss of mating ability in *MAT*a strains in a patch mating assay (Fig. 12A). This was confirmed by quantitative assays, which revealed a strong reduction of the relative mating ability of the *rap*⁻, *acs*⁻ and *D2*⁻ strains compared to the *HML-SS* ΔI strain with wild-type silencer elements (Fig. 12B). This showed that all three elements within the truncated 79-bp *HML*-E silencer were not only sufficient, but also essential for *HML* silencing.

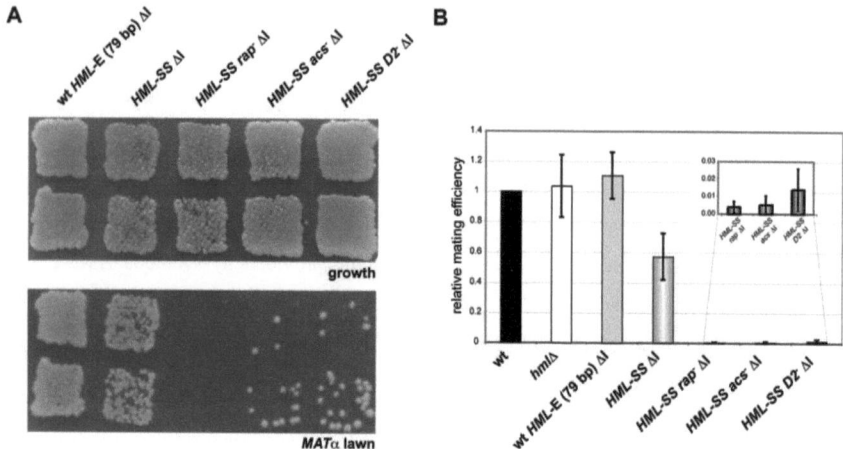

Fig. 12 Characterization of synthetic *HML*-E silencer constructs
Mutation of individual *HML*-E elements in the synthetic silencer caused *HML* derepression. (A) Mutations of the Rap1 (AEY4408) and ORC (AEY4410) binding sites and the D2 element (AEY4412) caused a loss of *HML* silencing as measured by a loss in mating ability. The figure shows a patch mating assay of *MAT*α yeast strains with genomically integrated *HML*-E alleles. YPD served as growth control. (B) Mating efficiency of *MAT*a strains carrying the indicated *HML* alleles was measured in a quantitative mating assay and normalized to the mating efficiency of a wt *HML* strain (AEY2). Bars for the relative mating efficiency of binding site mutants are enlarged in the inset. Error bars represent the standard deviations of three individual experiments.

3.1.3 Mutations *in trans* caused a reduction in silencing by the minimal *HML*-E silencer

Since mutations of the individual silencer domains of *HML-SS* ΔI silencer caused a loss of *HML* silencing, it was next analysed whether mutation or deletion of the genes encoding the respective binding proteins lead to a similar loss of silencing. To this end, strains were constructed which combined *HML-SS* ΔI with mutations in *RAP1*, ORC (because the genes are essential) or with the deletion of *SUM1*, and *HML* silencing was tested by measuring the mating ability of *MAT*a strains (Fig. 13 A-C). Importantly, the *rap1-12* mutation (Sussel & Shore, 1991) in combination with *HML-SS* ΔI caused a complete loss of silencing (Fig. 13A). The quantitative analysis showed that the relative mating ability of this strain was comparable to that of the *HML-SS rap⁻* ΔI strain (Fig. 13D). These data supported an essential role for Rap1 in *HML* silencing (Boscheron et al., 1996).

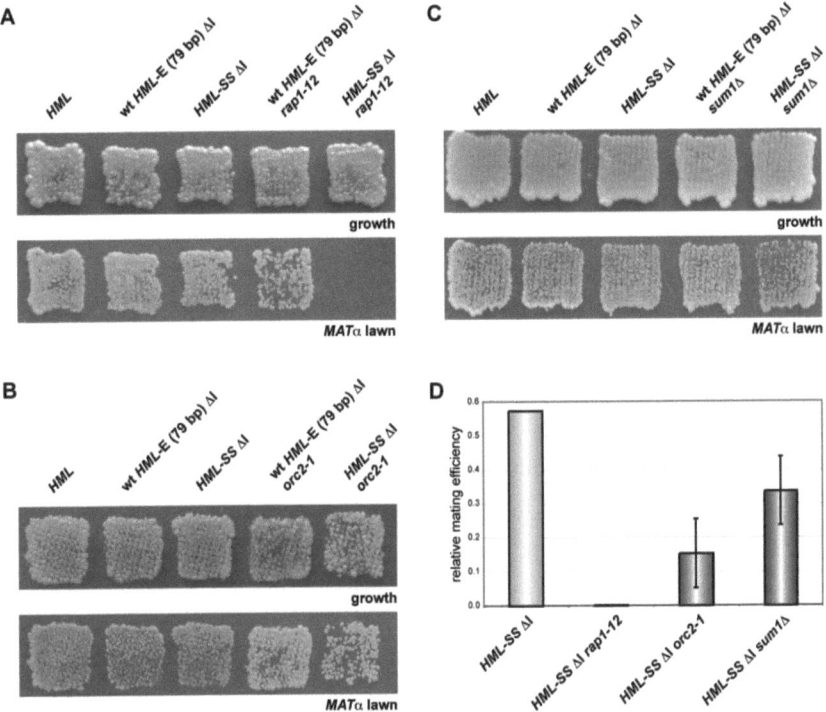

Fig. 13 Synthetic *HML*-E was sensitized for mutations in *RAP1* and ORC and for the deletion of *SUM1* (A-C) The mating ability of *MATa HML-SS* ΔI strains with *rap1-12* (AEY4464) (A), *orc2-1* (AEY4486) (B) or with *sum1*Δ (AEY4428) (C) was compared to that of corresponding wild-type strains with the indicated *HML* alleles (AEY2 / AEY4404 / AEY4406). (D) Quantitative mating efficiencies of *MATa* strains with the indicated genotypes. Error bars represent the standard deviations of three individual experiments.

Surprisingly, the *orc2-1* mutation within the ORC complex (Foss *et al.*, 1993) in combination with the *HML-SS* ΔI silencer displayed a much weaker silencing defect than the *HML-SS acs⁻* ΔI allele (Fig. 12A and Fig. 13A). The relative mating ability of this strain was approximately 10 % of the *HML-SS* ΔI control, but several-fold higher than that of the *HML-SS* ΔI *acs⁻* strain (Fig. 13B). This was surprising, because silencing by a synthetic silencer at the other *HM* locus, *HMR*, is sensitive to *orc* mutations (Fox *et al.*, 1995). The *orc5-1* mutation (Loo *et al.*, 1995a) was also tested in the *HML-SS* ΔI strain, but it also did not enhance *HML* derepression (data not shown). This indicated that for unknown reasons (see Discussion), the ACS of *HML-SS* was not sensitive to the *orc2-1* and *orc5-1* mutations. Similarly, the absence of Sum1, which has been shown to bind to the D2 element (Irlbacher *et al.*, 2005), caused only a slight reduction in silencing of *HML-SS* ΔI. The mating ability of a *MATa HML-SS* ΔI *sum1*Δ strain was reduced to approximately 50 % of that of a strain with the minimal *HML* silencer alone (Fig. 13C), and the effect was much less pronounced than for a D2 element mutation within the *HML-SS* ΔI variant (Fig. 12B and Fig. 13D). This showed that Sum1 had some effect on silencing, but further suggested that other factors are involved in silencing via the D2 element. Notably, as a *sum1*Δ was shown to be synthetically lethal with *orc2-1* (Irlbacher et al., 2005), it was not possible to analyse whether this double mutation would cause an additive effect on silencing with *HML-SS* ΔI.

To investigate the role of *in trans orc2-1* and *sum1*Δ mutations in *HML* silencing in more detail, an additional experiment was performed. Therefore, the *MATa1* promoter was deleted in strains carrying the synthetic, minimal *HML*-E silencer (*mat*Δ *HML-SS* ΔI). With wild-type HM silencers, *mat*Δ cells mate as **a** cells and in this context *HML* derepression would enable the cells also to grow on a *MAT***a** lawn. It was shown previously, that *mat*Δ *hmr*Δ strains were able to mate weakly as α cells when carrying additional *orc2-1* or *orc5-1* mutations (Loo et al., 1995a). As expected, a *mat*Δ *HML-SS* ΔI strain was able to grow on a *MAT*α lawn and less efficiently on a *MAT***a** lawn (Fig. 14). The deletion of *SUM1* impaired **a**-mating significantly, whereas it improved α-mating (Fig. 14) probably indicating enhanced *HML-SS* ΔI derepression by *sum1*Δ. While the reduction of **a**-mating ability was even stronger in an *orc2-1* strain, surprisingly, those cells did not show α-mating on a *MAT***a** lawn. This suggested that ORC is required for expression of α-information from *HML-SS* ΔI. To conclude, this additional experiment verified the enhancing *HML-SS* ΔI derepression effect of the *orc2-1* and *sum1*Δ mutations.

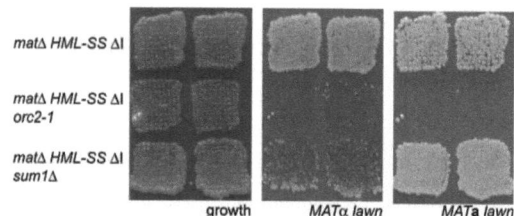

Fig. 14 *sum1*Δ **and** *orc2-1* **increased** *HML-SS* ΔI **derepression in** *mat*Δ **cells**
The **a**- and α–mating abilities of *mat*Δ *HML-SS* ΔI strains with *orc2-1* or with *sum1*Δ were compared to that of a corresponding *ORC2 SUM1* wild-type strain. YPD served as growth control.

3.1.4 Sir1 and Dot1 were required for silencing of *HML-SS* ΔI

Although Sir1 is required for full silencing of both *HML* and *HMR*, *sir1*Δ strains still show substantial silencing at both loci (Pillus & Rine, 1989; Rine & Herskowitz, 1987). However, *sir1Δ* causes complete derepression of *HMR* when controlled by synthetic *HMR*-E (Gardner et al., 1999). Also, *dot1Δ* only causes derepression of *HML* when silencing is previously compromised by *sir1Δ* (van Welsem et al., 2008). Since synthetic *HML*-E constitutes a sensitized silencer, this suggested that *sir1Δ* and *dot1Δ* on their own might be able to derepress *HML-SS* ΔI. Therefore, it was investigated whether *sir1Δ* or *dot1Δ* were capable of disrupting silencing in strains with the synthetic *HML-SS* ΔI allele. Significantly, both *sir1Δ* and *dot1Δ* caused a complete loss of mating ability, indicating a complete derepression of *HML-SS* ΔI (Fig. 15). This showed that the minimal *HML* silencer sensitized *HML* silencing to mutations in *SIR1* and *DOT1*.

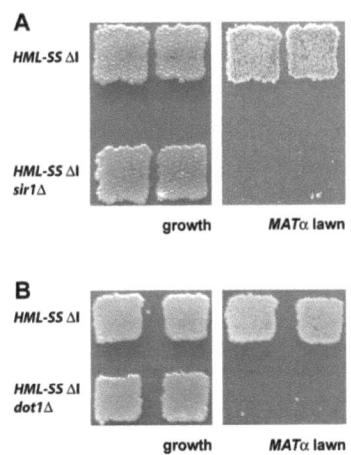

Fig. 15 Sir1 and Dot1 were essential for silencing of *HML-SS* ΔI
MATa HML-SS ΔI *sir1Δ* (AEY4538) (A) and *MATa HML-SS* ΔI *dot1Δ* (AEY4805) (B) cells were tested for their ability to mate with a *MATα* tester strain in a patch mating assay. YPD served as growth control.

3.1.5 Genetic screen to search for additional novel factors influencing *HML* silencing

The synthetic minimal *HML* silencer was used in a genetic screen with the *S. cerevisiae* deletion library to find new factors that are required for *HML* silencing but so far have not been identified due to the high redundancy of natural *HML*-E. To this end, a genetic screen previously designed to identify synthetic lethal interactions (Tong et al., 2001) was modified in order to obtain *MATa* cells with the *HML-SS* ΔI allele that carried a gene disruption from the deletion library (see 2.10.3). The first strain used for this approach (AEY3054, (Tong et al., 2004)) allowed – theoretically – with a *can1Δ::MFA1pr-HIS3-MFα1pr-LEU2* construct to select for haploid *MATa* cells on medium with the toxin canavanine and lacking histidine and for haploid *MATα* cells on canavanine medium lacking leucine. However, control experiments revealed that the selection for *MATa HML-SS* ΔI cells was insufficient so that a mixture of haploid *MATa* and *MATα* cells as well as diploids grew on the last minimal medium plates prior to the mating test on a *MATα* lawn. Therefore, the strain background was unsuitable for the sensitive silencing assay. Due to that, a second strain (AEY4573) was used that allowed a more stringent selection for haploid *MATa* cells in genetic screens ((Tong & Boone, 2007); personal communication Ehrenhofer-Murray and van Leeuwen 2009) (Fig. 16). A patch mating assay was performed to analyse *HML-SS* ΔI silencing of strains carrying individual gene deletions from the deletion library.

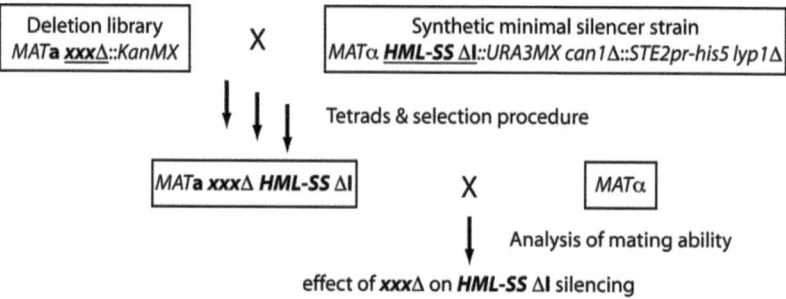

Fig. 16 Illustration of the experimental procedure to identify novel factors that influence silencing by the synthetic *HML*-E silencer
The *S. cerevisiae MATa* deletion library was used for a genetic cross with the *MATα* strain carrying *HML-SS* ΔI marked with *URA3MX*. The experimental procedure selected for haploid cells with the *HML-SS* ΔI allele that carried the respective gene deletion that was marked with *KanMX*. A mating assay with a *MATα* tester strain was used to measure silencing of *HML-SS* ΔI.

After a first screen with the complete deletion library, 264 primary candidates that showed reduced mating ability were selected for a second, independent screening procedure. However, due to differences in the laboratory's single cryotube collection and the 96-microplate deletion library, only 248 candidates were tested in the re-array. To screen the secondary candidates, two independent experimental approaches were used. First, the genetic screen was repeated with the primary candidates. Thus, 17 strains were excluded from the candidate list, because the *HML-SS* ΔI segregants carrying a gene deletion showed strong mating ability, whereas for others, a reduced mating ability was confirmed (for an example see Fig. 17).

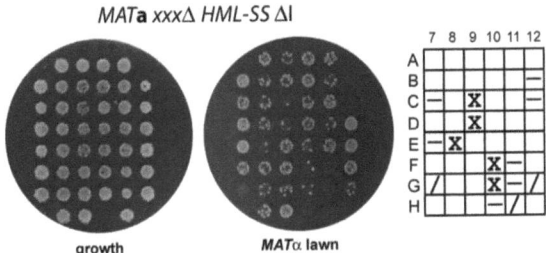

Fig. 17 Re-evaluation of *HML-SS* ΔI defects in candidate strains
An exemplary final mating test plate after a repetition of the genetic screen with the *S. cerevisiae* deletion library is shown. Primary candidates carrying the synthetic *HML-SS* silencer are classified into different groups of deletion strains putatively affecting *HML* silencing. "X" indicates secondary candidates that showed a reduced mating ability and which were used for tetrad dissection to test *HML-SS* ΔI silencing. Deletion strains with wild-type *HML*-E silencer that failed to form diploids with on a *MAT*α tester strain are marked with a diagonal line (/), those that did not pass the selection procedure of the genetic screen are marked with a horizontal line (—). YPD served as growth control.

Secondly, all primary candidates were transformed with a *HML-SS* ΔI-carrying plasmid (pAE1484) and their mating ability with and without the plasmid was tested. Strains that displayed reduced mating only in the presence of the plasmid were considered for a top list of factors putatively influencing *HML* silencing (Fig. 18).

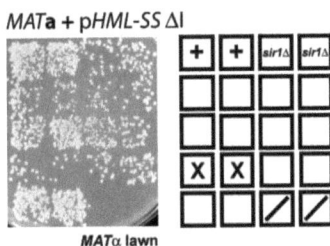

Fig. 18 Secondary screen of selected candidate strains for *HML-SS* ΔI silencing defects
Strains from the deletion library transformed with an *HML-SS* ΔI plasmid (pAE1484) were tested for the mating ability on a *MAT*α lawn. *sir1*Δ served as a control for impaired *HML-SS* ΔI silencing, a strain that showed normal mating in the primary screen was used as a positive mating control (+). Strains that were removed from the candidate list after the mating test of deletion library strains (Fig. 19) are marked with a diagonal line (/), strains that showed reduced mating ability with the *HML-SS* ΔI plasmid were considered as secondary candidates and marked with "X".

In order to eliminate those primary candidates whose reduced mating ability was *HML-SS* ΔI-independent, all candidates were tested for mating defects caused by the respective gene disruption alone (Fig. 19). After this, test twelve strains were excluded from the primary list. Taken together, both experiments led to 219 secondary candidates (Table 8) and two separate top lists (Tables 9 and 10) of deletions that were candidates for causing the strongest *HML* derepression with the minimal *HML* silencer. One top list (Table 9) contains those gene deletions that caused the strongest mating defect in an *HML-SS* ΔI strain in the re-array of the genetic screen (for an example see Fig. 17), and the other top list (Table 10) contains those strains that showed the strongest effect when carrying a *HML-SS* ΔI plasmid (for an example see Fig. 18).

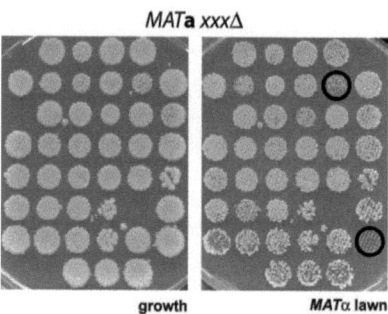

Fig. 19 Mating defects of some strains from the deletion library with wild-type *HML*-E allele
Primary candidate strains from the *S. cerevisiae* deletion library were tested for mating ability. An exemplary plate that contained strains displaying mating defects with wild-type *HML*-E (highlighted with black circles) is shown. YPD served as growth control.

Table 8: Complete list of secondary screen candidates

ORF	Gene[1]	ORF	Gene[1]	ORF	Gene[1]	ORF	Gene[1]	ORF	Gene[1]
YAL020C	ATS1	YDR137W	RGP1	YLR384C	IKI3	YHR004C	NEM1	YIL125W	KGD1
YAL010C	MDM10	YDR354W	TRP4	YLR388W	RPS29A	YHR026W	PPA1	YIL149C	MLP2
YAL002W	VPS8	YDR363W	ESC2	YLR417W	VPS36	YHR059W	FYV4	YIL163C	Dubious
YAR044W	SWH1	YDR369C	XRS2	YLR418C	CDC73	YOL158C	ENB1	YDR007W	TRP1
YLL038C	ENT4	YDR389W	SAC7	YDR150W	NUM1	YJR102C	VPS25	YFR011C	AIM13
YLR021W	IRC25	YEL046C	GLY1	YDR195W	REF2	YKR101W	SIR1	YNL056W	OCA2
YLR023C	IZH3	YER019W	ISC1	YPR179C	HDA3	YLR436C	ECM30	YCR084C	TUP1
YLR025W	SNF7	YER086W	ILV1	YDR535C	Dubious	YCL058C	FYV5	YGR155W	CYS4
YLR048W	RPS0B	YGR157W	CHO2	YBL060W	YEL1	YCL074W	pseudo	YML041C	VPS71
YLR069C	MEF1	YGR159C	NSR1	YGL024W	dubious	YGL211W	NCS6	YOR298C-A	MBF1
YLR074C	BUD20	YGR178C	PBP1	YGL054C	ERV14	YGL212W	VAM7	YPL183W-A	RTC6
YLR087C	CSF1	YHR073W	OSH3	YGL064C	MRH4	YGL213C	SKI8	YOR302W	CPA1
YLR091W	GEP5	YHR207C	SET5	YGL072C	dubious	YGL218W	dubious	YDL115C	IWR1
YLR104W	LCL2	YCL008C	STP22	YGL084C	GUP1	YGL252C	RTG2	YJL101C	GSH1
YML087C	AIM33	YCL030C	HIS4	YNL238W	KEX2	YGL260W	unchar	YBR134W	Dubious
YML024W	RPS17A	YCL062W	VAC17	YNL215W	IES2	YPL090C	RPS6A	YDL046W	NPC2
YML016C	PPZ1	YCL064C	CHA1	YNL148C	ALF1	YPL064C	CWC27	YDL081C	RPP1A
YML002W	unchar.	YLR182W	SWI6	YKL191W	DPH2	YCL075W	pseudo	YDR455C	Dubious
YMR032W	HOF1	YKL003C	MRP17	YKL212W	SAC1	YJR097W	JJJ3	YGL105W	ARC1
YMR153C-A	dubious	YKL031W	dubious	YKL213C	DOA1	YJR104C	SOD1	YER155C	BEM2
YMR178W	unchar.	YKL009W	MRT4	YKL216W	URA1	YJR109C	CPA2	YJR055W	HIT1
YMR228W	MTF1	YKL067W	YNK1	YKR020W	VPS51	YJR139C	HOM6	YIL094C	LYS12
YNL335W	DDI3	YKL110C	KTI12	YKR042W	UTH1	YDL191W	RPL35A	YDL047W	SIT4
YOR001W	RRP6	YKL149C	DBR1	YDR295C	HDA2	YDL198C	GGC1	YER014W	HEM14
YOR038C	HIR2	YKL169C	dubious	YDR298C	ATP5	YDL232W	OST4	YEL044W	IES6
YOR068C	VAM10	YKL170W	MRPL38	YDR323C	PEP7	YDR034C	LYS14	YER087W	AIM10
YOR070C	GYP1	YKL176C	LST4	YIL052C	RPL34B	YBR285W	unchar.	YLR226W	BUR2
YOR302W	YOR302W	YKL190W	CNB1	YFR009W	GCN20	YCR065W	HCM1	YCR063W	BUD31
YOR359W	VTS1	YGR056W	RSC1	YKR035W-A	DID2	YCR081W	SRB8	YGR257C	MTM1
YOL002C	IZH2	YGR061C	ADE6	YKR082W	NUP133	YJR063W	RPA12	YBL093C	ROX3
YOL024W	YOL024W	YGR064C	dubious	YOL111C	MDY2	YDL116W	NUP84	YBR112C	CYC8
YOL063C	CRT10	YGR101W	PCP1	YOL138C	RTC1	YDL151C	BUD30	YNL307C	MCK1
YPL259C	APM1	YOR141C	ARP8	YHR064C	SSZ1	YDL172C	dubious	YLR240W	VPS34
YPL178W	CBC2	YOR188W	MSB1	YNL250W	RAD50	YDL173W	PAR32	YLR244C	MAP1
YPL173W	MRPL40	YOR193W	PEX27	YNL252C	MRPL17	YNL021W	HDA1	YLR396C	VPS33
YPL129W	TAF14	YOR211C	MGM1	YOR096W	RPS7A	YNL032W	SIW14	YGL020C	GET1
YPL106C	SSE1	YJL204C	RCY1	YMR111C	unchar	YDR440W	DOT1	YGR262C	BUD32
YBR185C	MBA1	YJL189W	RPL39	YMR116C	ASC2	YNR010W	CSE2	YGL218W	Dubious
YBR205W	KTR3	YJL188C	BUD19	YPR070W	MED1	YNR037C	RSM19	YOR369C	RPS12
YBR226C	YBR226C	YJL183W	MNN11	YPR074C	TKL1	YBR036C	CSG2	YDL160C	DHH1
YBR245C	ISW1	YJL134W	LCB3	YJL120W	dubious	YBR037C	SCO1	YDR432W	NPL3
YBR246W	RRT2	YJL130C	URA2	YJL088W	ARG3	YBR041W	FAT1	YDR297W	SUR2
YDR129C	SAC6	YJL129C	TRK1	YJL062W	LAS21	YBR047W	FMP23	YBR082C	UBC4
YDR136C	VPS61	YLR372W	SUR4	YJL036W	SNX4	YNL139C	THO2		

[1] Standard gene names are given as annotated in the *Saccharomyces* Genome Database to the corresponding ORF. Pseudogenes, uncharacterized or dubious ORFs are indicated.

Tables 9: Top list of gene disruptions with reduced mating ability in a genetic screen re-array

ORF	Gene	Position in deletion library
YMR228W	MTF1	4G5
YOR038C	HIR2	6E7
YPL173W	MRPL40	9D6
YDR129C	SAC6	11E4
YLR182W	SWI6	17B10
YJL188C	BUD19	21C4
YLR417W	VPS36	22D10
YKR085C	CDC73	22D11
YBR041W	FAT1	42B11
YER155C	YME1	49G11

Tables 10: Top list of gene disruptions with reduced mating ability with HML-SS ΔI plasmid

ORF	Gene	Position in deletion library
YBR037C	SCO1	42B9
YOR369C	RPS12	71A9
YBR205W	KTR3	10C9
YBR226C	YBR226C	10E5
YBR246W	RRT2	10F8
YDR389W	SAC7	12A9
YNL139C	THO2	42F2
YKR042W	UTH1	29F2
YKL169C	YKL169C	18H3
YHR073W	OSH3	15B9

Although not included in the two lists of top candidates, three additional gene deletions were chosen (Table 11) for further analyses, since the respective genes were annotated to function in chromatin remodelling (*ISW1*), histone deacetylation and telomere maintenance (*HDA2*) and transcriptional regulation (*CYC8*).

Table 11: List of additionally tested secondary candidates

ORF	Gene	Position in deletion library
YBR245C	ISW1	10F7
YDR295C	HDA2	30C7
YKR085C	MRPL20	71C5*

* Barcode analysis revealed *mrpl20Δ* instead of *cyc8Δ* at position 71C5 (For explanation see next section)

Top-scoring candidates did not cause derepression of *HML-SS* ΔI

In a next step it was sought to verify the derepressing effect of selected gene deletions on *HML-SS* ΔI silencing by tetrad analysis. The 23 top-scoring candidates (Tables 9-11) were chosen for this purpose to validate a potential influence on *HML-SS* ΔI. The mating ability of four *HML-SS* ΔI strains carrying the respective gene deletion generated by tetrad dissection was compared to *HML-SS* ΔI and *xxx*Δ single mutants in a patch mating assay. However, although the strains of each type had the same genotype according to marker analysis, the patches of all tested mutants showed a variable mating ability on a *MAT*α lawn (Fig. 20). This could be explained by genetic heterogeneity due to additional unknown and unmarked mutations of the deletion strains, which might affect mating ability but were not traceable during tetrad analysis.

Fig. 20 Variability of silencing defects in genetically identical segregants
*MAT*a gene deletion (here exemplary *mtf1*Δ) cells with the synthetic *HML-SS* ΔI allele were tested for their ability to mate with a *MAT*α tester strain in comparison to strains with *HML-SS* ΔI and wild-type gene (*MTF1*) or gene deletion (*mtf1*Δ) and wild-type *HML*-E allele in a patch mating assay. YPD served as growth control.

HIR2 is an example for a candidate gene that might influence *HML* silencing, since the mating ability of a *HML-SS* ΔI *hir2*Δ double mutant was significantly reduced after the repetition of the genetic screen as compared to a strain with the minimal *HML*-E silencer alone. Eight *HML-SS* ΔI *hir2*Δ and eight *HML-SS* ΔI *HIR2* mutant strains from tetrad dissection were tested in a patch mating assay on a *MAT*α lawn. However, this experiment led to a very heterogenous picture, since the different strains with the same genomic markers showed a varying mating ability (Fig. 21A). Some *HML-SS* ΔI *hir2*Δ double mutant patches exhibited a clear reduction in mating ability, whereas others were indistinguishable from the single mutant patches, thus making a general role of Hir2 in *HML* silencing unlikely. Nevertheless, additional experiments were performed to validate Hir2 as a factor influencing *HML* silencing. The identity of the *HIR2* deletion of the candidate was controlled by a PCR-based barcode analysis (Baudin *et al.*, 1993; Wach *et al.*, 1994). Here, the flanking region of the *KanMX* cassette, which varies in all deletion library strains and is unique to each gene deletion, was sequenced and confirmed the *hir2Δ::KanMX* mutation. Furthermore, to rule out side effects of the deletion library or screening strain (AEY4629) background, *HIR2* was disrupted in an *HML-SS* ΔI w303 strain (AEY4752). In this strain background *hir2Δ* did not cause derepression of *HML-SS* ΔI (Fig. 21B). This led to the conclusion that *HML-SS* ΔI silencing was Hir2-independent.

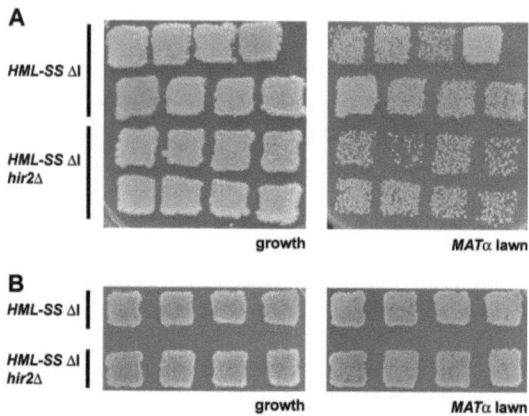

Fig. 21 *hir2Δ* did not affect silencing of *HML-SS* ΔI
*MAT*a *HML-SS* ΔI *hir2Δ* cells with the genetic screen (AEY4754) (A) or W303 background (AEY4807) (B) were tested for their ability to mate with a *MAT*α tester strain in a patch mating assay and compared to *HML-SS* ΔI mutants with wild-type *HIR2*. YPD served as growth control.

The same detailed analysis was performed for another factor, Mrpl20, which also showed no reduced mating in an *HML-SS* ΔI w303 strain upon deletion (Fig. 22). Initially, due to its position in the deletion library (71C5) this candidate supposedly was thought a *cyc8Δ* strain. However, barcode analysis provided evidence that it was instead an *mrpl20Δ* strain (originally position 71C6). This observation demonstrated a mismatch between the content of the deletion library microplates and the given gene deletion list and indicated the requirement for the barcode analysis of the candidate strains to identify the respective gene deletion.

To summarize, although the genetic screen seemed to be a powerful tool to identify novel factors that contribute to *HML* silencing, apparent genetic differences of the deletion library strains apart from the respective gene deletion caused heterogenous silencing effects and precluded identifying such factors.

Fig. 22 mrpl20Δ did not cause HML-SS ΔI derepression
MATa HML-SS ΔI mrpl20Δ W303 cells were tested for their ability to mate with a MATα tester strain in a patch mating assay. YPD served as growth control.

In independent experiments, a candidate gene approach was used in parallel to the validation of the candidates from the genetic screen to analyse other gene deletions for their ability to cause derepression of HML-SS ΔI. Sas2 and Asf1 are other factors that have previously been shown to cause HML derepression upon deletion in a sir1Δ background (Ehrenhofer-Murray et al., 1997; Meijsing & Ehrenhofer-Murray, 2001; Osada et al., 2001; Reifsnyder et al., 1996). However, asf1Δ and sas2Δ did not impair silencing of HML-SS ΔI (Fig. 23 A / data not shown). Furthermore, the effect of a HIR1 deletion in HML-SS ΔI strain was analysed, since hir1Δ leads to a loss of HML silencing in a triple mutant strain with sir1Δ and cac1Δ (Meijsing & Ehrenhofer-Murray, 2001). Additionally Hir1, like Hir2, which appeared as a candidate in the genetic screen with the deletion library, is another component of the HIR nucleosome assembly complex (Sharp et al., 2001). Similarly, the absence of Hir1 did not affect HML-SS ΔI (Fig. 23B). Taken together, these results showed that the synthetic HML-E silencer sensitized silencing to some silencing factors, but that the sensitization was different from that caused by the absence of Sir1.

Fig. 23 HIR complex components did not influence *HML-SS* ΔI silencing
*MAT*a *HML-SS* ΔI *asf1*Δ (A) and *MAT*a *HML-SS* ΔI *hir1*Δ (B) W303 cells were tested for their ability to mate with a *MAT*α tester strain in a patch mating assay. YPD served as growth control.

The results presented so far indicated an involvement of several factors in the regulation of *HML* silencing. With the help of mutations *in cis* and *in trans* it was shown that not only Rap1 but also ORC and Sum1 contributed to the silencing of *HML-SS* ΔI. It has to be noted that *HML*-E not only serves as a silencer element of *HML*, but it also has origin functions (ARS301) on a plasmid (Vujcic *et al.*, 1999). Although it is capable of confering replication, in the genomic context it is a passive origin that is replicated by neighbouring origins located adjacent to the chromosomal *HML* region. Nevertheless, it shares essential features such as ACS or binding sites for auxiliary factors with active *S. cerevisiae* origins. This made it interesting to investigate whether common elements that have been shown to influence *HML-SS* ΔI silencing also contribute to the process of DNA replication at certain origins.

3.2 Histone deacetylation affected replication initiation at a subset of origins

ORC exhibits a dual role in *HML* silencing as the recruitment complex for the Sir proteins and in origin function, because it nucleates the assembly of the pre-replicative complex to initiate DNA replication (reviewed in (Rehman & Yankulov, 2009)). However, ORC is not the only structural component that is present in these distinct pathways. Besides this, Sum1 does not only bind to the D2 element at *HML*-E, but is also part of a histone deacetylase complex. Therefore, the influence of Sum1 and ORC on origin activity at several genomic origins was analysed.

3.2.1 *hst1Δ* and *rfm1Δ* were synthetically lethal with an *orc2-1* mutation

Sum1 interacts with Hst1 via Rfm1 (McCord *et al.*, 2003), and this Sum1/Rfm1/Hst1 complex represses a number of midsporulation genes. Therefore, the question arose whether this complex was also involved in Sum1's initiation function, which is reflected in the observation that an *orc2-1* mutation is synthetically lethal in combination with *sum1Δ* (Irlbacher *et al.*, 2005; Suter *et al.*, 2004). To this end, it was investigated whether *hst1Δ* and *rfm1Δ* were also lethal with *orc2-1*. Significantly, an *orc2-1* strain with *hst1Δ* was only able to lose an *URA3*-marked *ORC2* plasmid on counterselective medium (5-FOA) if the strain had previously been provided with an *ORC2*-carrying plasmid with a different selection marker, but not with a vector control (Fig. 24), which was in agreement with previous work (Suter *et al.*, 2004). Additionally, it was found that *rfm1Δ* was synthetically lethal with *orc2-1* (Fig. 24), indicating that that the whole Sum1/Rfm1/Hst1 complex was involved in the initiation function of Sum1.

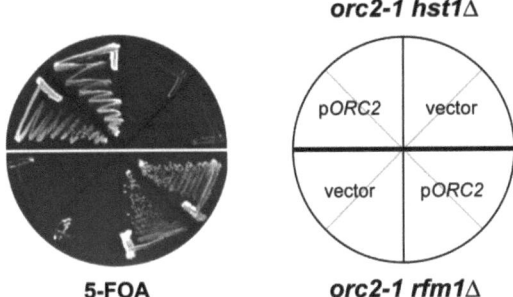

Fig. 24 *hst1Δ* and *rfm1Δ* showed synthetic lethality with *orc2-1*
An *orc2-1 hst1Δ* (AEY3941) and an *orc2-1 rfm1Δ* (AEY3940) strain carrying p*URA3-ORC2* (pAE1316) were tested on 5-FOA medium for their ability to lose the *ORC2* plasmid when additionally provided with a *LEU2*-marked *ORC2* plasmid (pAE1315) or an empty *LEU2* vector (pRS315). Strains were incubated for three days at 23°C.

3.2.2 ARS activity of selected origins depended on Sum1 and Hst1

In previous work, bioinformatics analysis of genome-wide binding studies was used to identify regions in the *Saccharomyces cerevisiae* genome that are bound by both ORC and Sum1 and thus are good candidates for origins of replication that are regulated by Sum1 (Fig. 25, (Irlbacher et al., 2005)). This analysis revealed eight regions that showed binding of both ORC and Sum1, and it was shown previously for three of these fragments that they are ARS elements and require Sum1 for full initiation capacity (Irlbacher et al., 2005). To further validate these putative Sum1-regulated origins, it remained to be determined whether the genes downstream of the Sum1 binding sites were repressed by Sum1 and Hst1. To this end, existing microarray data for the expression of these genes in $sum1\Delta$ and $hst1\Delta$ cells were queried (Bedalov et al., 2003). This analysis showed that six of the genes were upregulated upon deletion of *SUM1* or *HST1* (Table 12), indicating that the localization of Sum1 to these fragments likely recruits Hst1 to repress the neighbouring gene through histone deacetylation. In the other two cases, Sum1 may solely act as a replication factor, because it does not repress the gene next door.

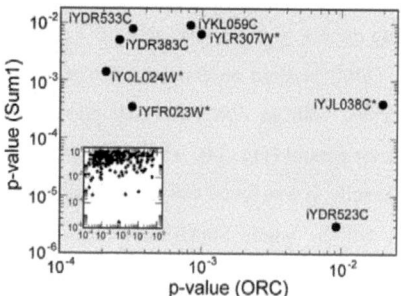

Fig. 25 Search for intergenic regions that bind both Sum1 and ORC
Plot of p-values for Sum1 binding, $P < 0.01$ (Lee et al., 2002) versus ORC binding, $P < 0.05$ (Wyrick et al., 2001). All data points are enlarged in the inset (Irlbacher et al., 2005).

Table 12: Gene expression change in $sum1\Delta$ and $hst1\Delta$ strains compared to wild-type

ARS	Intergenic region	Gene	$sum1\Delta$ / WT[*]	$hst1\Delta$ / WT[*]
433	iYDR383C	NKP1	1.0	1.0
446	iYDR523C	SPS1	21.6	12.6
447	iYDR533C	HSP31	2.3	1.6
607	iYFR023W	PES4	3.4	2.0
1013	iYJL038C	YJL038C	2.8	2.0
1109	iYKL059C	MPE1	0.8	1.0
1223	iYLR307W	CDA1	25.5	6.2
1511	iYOL024W	YOL024W	5.1	1.5

[*] data from (Bedalov et al., 2003)

It was asked next whether the ARS activity of these putative origins was regulated by Sum1 and Hst1. To test this, *in vivo* plasmid maintenance of these regions was measured in wild-type, *sum1Δ* and *hst1Δ* strains.

An earlier study had shown that ARS1013, ARS1223 and ARS1511 depended on Sum1 for full initiation (Irlbacher *et al.*, 2005). Here, it was found that *sum1Δ* and *hst1Δ* strains with *CEN4-URA3* plasmids containing ARS446, ARS607, ARS1013, ARS1109, ARS1223 or ARS1511 as the sole origin displayed a significantly higher plasmid loss rate than the corresponding wild-type strain (Fig. 26A), showing that both Sum1 and Hst1 were necessary for the ARS activity of these origins. Interestingly, most origins showed a stronger dependence on Sum1 than on Hst1, reflecting the observation that many genes show stronger derepression by *sum1Δ* than by *hst1Δ* (Table 12, (Bedalov *et al.*, 2003)). The effect of *sum1Δ* and *hst1Δ* on ARS activity was specific to Sum1-bound origins, because a control origin that is not bound by Sum1, ARSH4, did not show an increased plasmid loss rate (Irlbacher *et al.*, 2005).

In several cases, the plasmid loss was too high to measure a plasmid loss rate, because primary transformants failed to grow upon restreaking. This was the case for ARS446 in a *sum1Δ* background and for ARS1013 in *sum1Δ* and *hst1Δ* strains (Fig. 26B).

One origin, ARS433, displayed dependence on Sum1, but not Hst1 (Fig. 26A). This reflected the fact that the neighbouring gene, *NKP1*, was not repressed by Hst1 or Sum1 (Table 12) and suggested that this origin was regulated by Sum1 independently of the Sum1/Rfm1/Hst1 complex.

The analysis was further extended to the ARS606 origin that had previously been identified as a Sum1-regulated origin by searching for co-occurrences of an ACS and a Sum1 consensus-binding site (Irlbacher *et al.*, 2005). ARS606 was highly unstable in *sum1Δ* as well as in *hst1Δ* strains (Fig. 26B), thus precluding the measurement of plasmid loss rates and showing that this Sum1-regulated origin also depended upon Hst1.

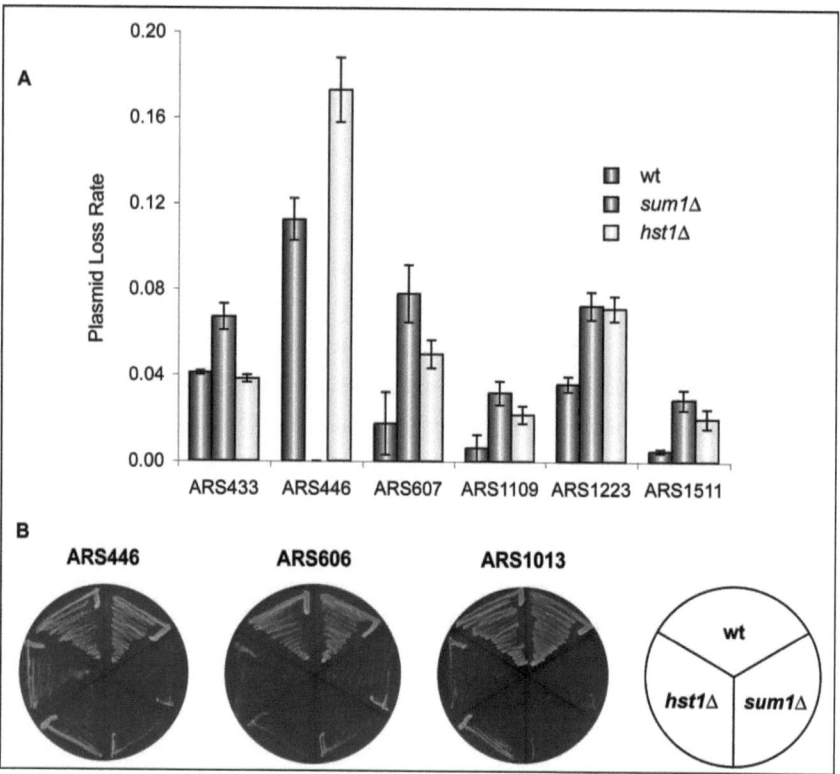

Fig. 26 Sum1 and Hst1 were necessarry for ARS activity of selected origins
(A) Plasmid loss rates were measured in a wild-type (wt, W303, purple), a *sum1Δ* (AEY3358, red) and an *hst1Δ* (AEY1499,) strain. Strains carried *CEN4-URA3* plasmids with ARS433 (pAE1240), ARS446 (pAE1250), ARS607 (pAE1242), ARS1109 (pAE1243), ARS1223 (pAE1130) or ARS1511 (pAE1135) as their sole origin. The loss rates are the average of three independent determinations. No loss rate could be determined for ARS446 in the *sum1Δ* strain. (B) Primary transformants of wt, *sum1Δ* and *hst1Δ* strains carrying *CEN4-URA3* plasmids with ARS446 (pAE1250), ARS606 (pAE1126) or ARS1013 (pAE1081) were streaked on minimal plates lacking uracil and incubated for three days at 30°C.

In contrast to other intergenic fragments, the region designated ARS447 was not capable of ARS activity, because wild-type and mutant strains transformed with ARS447 plasmids formed pinprick colonies that did not develop into viable cells after restreaking (data not shown). The ARS activity of a 0.5-kB as well as a 1.5-kB fragment comprising the putative ORC and Sum1-binding region was tested, but both failed to support autonomous replication. This was in agreement with the fact that this region is designated ARS447 by (Wyrick *et al.*, 2001), but not in the *Saccharomyces* Genome Database (SGD). A detailed description of the ARS fragments, the presence of ACS and Sum1 binding sites and their position relative to neighbouring genes is provided in Figure 27. (Fig. 27).

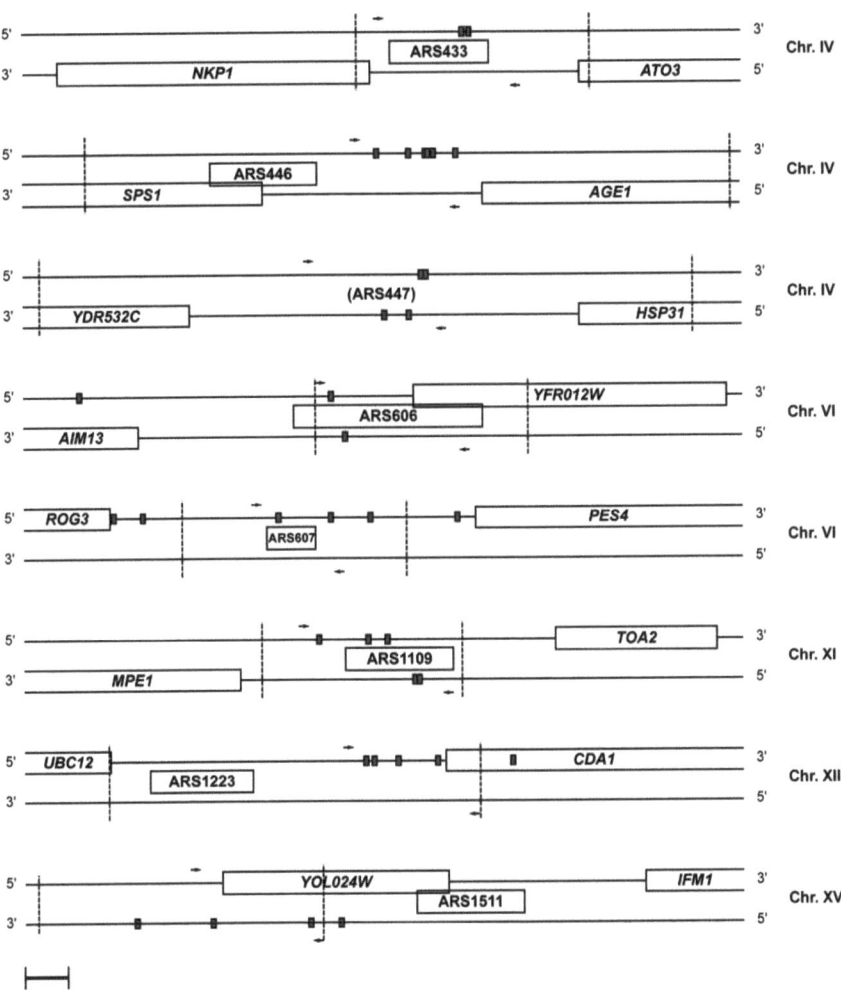

Fig. 27 Schematic representation of the ARS sequences analysed in this study
Genomic fragments used for plasmid loss assays are indicated by dashed, vertical lines. Black arrows represent the positions of the oligonucleotides used for quantitative real-time PCR for ChIP analysis. ARS sequences as annotated in the *Saccharomyces* genome database (SGD) are given by the marked boxes (ARS447, which is not annotated in SGD, is marked in brackets). The green and red rectangles show the position of the ARS and Sum1 consensus sequences with at least 10 of 11 matches on the Watson or Crick strand. Neighbouring genes are represented by (open or closed) boxes with the respective ORF or gene name.

Taken together, these data showed that seven origins that were bound by Sum1, required both Sum1 and Hst1 for full initiation activity, suggesting that Sum1 recruited the Sum1/Rfm1/Hst1 complex to these origins, and that histone deacetylation by Hst1 contributed to initiation function.

3.2.3 *sum1Δ* and *hst1Δ* caused increased histone H4 aceylation at selected replication origins

The dependence of origin function on the HDAC Hst1 suggested that histone deacetylation was necessary for efficient initiation activity. Therefore, it was determined whether acetylation levels at these origins increased in the absence of Sum1 or Hst1. *hst1Δ* has previously been shown to moderately increase H3 and H4 acetylation (Robert et al., 2004), but Hst1 specificity so far has not been determined. For this purpose chromatin immunoprecipitations (ChIP) with antibodies against different histone H4 acetyl-lysine residues were performed. Importantly, it was found that acetylation of H4 K5 was significantly increased at most ARS in the absence of Sum1 and of Hst1 (Fig. 28A). These results indicated that H4 K5 was a target for deacetylation by Hst1. In contrast, H4 K5 acetylation was not increased at a control region not bound by Sum1, ARSH4 (Fig. 28A). *CDC20* served as an additional control region that did not show a significant increase in acetylation with the used antibodies (Fig. 28).

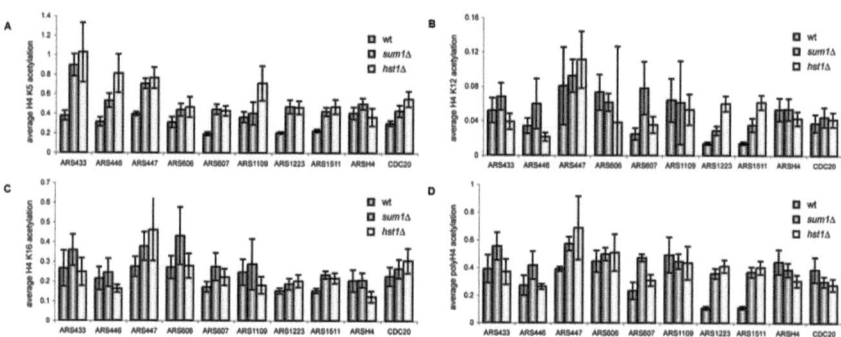

Fig. 28 *sum1Δ* and *hst1Δ* caused increased acetylation of H4 K5 at selected origins of replication
The amount of DNA from immunoprecipitated wild-type (wt, AEY2, purple), *sum1Δ* (AEY3358, red) and *hst1Δ* (AEY1499, blue) strains with anti-acetyl-histone H4 antibodies is shown relative to the input DNA. Quantitative qPCR was performed for eight selected origins and two controls (ARSH4 and CDC20). Error bars represent the standard deviation of six samples from two independent experiments. (A) anti-acetyl-histone H4 K5, (B) anti-acetyl-histone H4 K12, (C) anti-acetyl-histone H4 K16, (D) anti-acetyl-histone H4.

Furthermore, this observation suggested that Hst1 directly affected replication initiation at these origins by deacetylating histone H4 K5. There were three exceptions to this scenario. ARS433 plasmid maintenance was independent of Hst1, but ARS433 showed an Hst1-dependent increase in H4 K5 acetylation. Furthermore, Sum1 was required for full ARS activity of ARS606 and ARS1109, but apparently did not affect their acetylation state

(Fig. 28A). Thus, there seem to be scenarios where the relationship between Sum1, histone deacetylation and initiation is more complex.

In contrast to H4 K5, H4 K12 acetylation was not significantly increased at Sum1- and Hst1-dependent ARS elements. Whereas ARS1223 and ARS1511 showed a higher amount of H4 K12 acetylation in *hst1Δ* cells, the effect for ARS607 was only significant in the *sum1Δ* strain. The other origins showed no increase in acetylation (Fig. 28B). This suggested that H4 K12 was not a major target for deacetylation by Hst1, and that this site did not contribute to the regulation of initiation.

Similarly, the analysis presented here indicated that H4 K16 was not a general target of deacetylation by Hst1. Only ARS1223 and ARS1511 showed a significant higher H4 K16 acetylation level in both *sum1Δ* and *hst1Δ* yeast strains (Fig. 28C).

ChIP analysis with a poly-acetyl-H4 antibody showed overall increases in the acetylation levels at these origins in *sum1Δ* and *hst1Δ* strains. Whereas ARS1223 and ARS1511 showed a several fold higher acetylation, the effect was weaker for ARS447 and ARS607 and only significant for the *sum1Δ* strain. ARS433 and ARS446 showed no significant increase in H4 acetylation, and ARS606 and ARS1109 had the same state of acetylation in all three strains (Fig. 28D). This was consistent with the notion that H4 K5 was the main target of the histone deacetylase Hst1, whereas other histone H4 lysine residues were minor or no targets of Hst1.

To summarize, this analysis suggested that Sum1 regulated initiation at selected origins by recruiting Hst1, and thus histone deacetylation by the Sum1/Rfm1/Hst1/ HDAC complex, to these regions.

3.2.4 Changes in H4 acetylation caused defects in plasmid stability

The previous experiments raised the question whether increases in histone acetylation at origins by *sum1Δ* or *hst1Δ* were responsible for the loss of origin activity. To test this, it was asked whether the activity at these origins was decreased in a strain in which the acetylatable lysine residues of the H4 N-terminus (K5, 8, 12 and 16) were mutated to glutamine. These mutations mimic a constant acetylation state of the respective histone H4 residue as is the case in the absence of the HDAC Hst1. The effect of this mutation on plasmid stability of the two origins that were most strongly affected by Sum1 and Hst1, ARS606 and ARS1013, was analysed. Significantly, strains with ARS606 or ARS1013 plasmids showed a reduced growth rate in an H4 mutant strain as compared to wild-type (Fig. 29), indicating a loss of plasmid stability. In contrast, ARSH4, whose acetylation level did not alter in the absence of Hst1 or Sum1, did not show a reduced growth rate in the H4 mutant strain (Fig. 29). It has previously been reported that a simultaneous mutation of histone H4 K5, 8, 12, 16 to glutamine lengthens the cell cycle (Megee *et al.*, 1990). However, the observation that a difference in the growth ability of mutant cells compared to wild-type with the control plasmid (ARSH4) was not detected indicated that the growth differences with ARS606 and ARS1013 were specific to these origins and not due to a generalized growth defect of the H4 mutation.

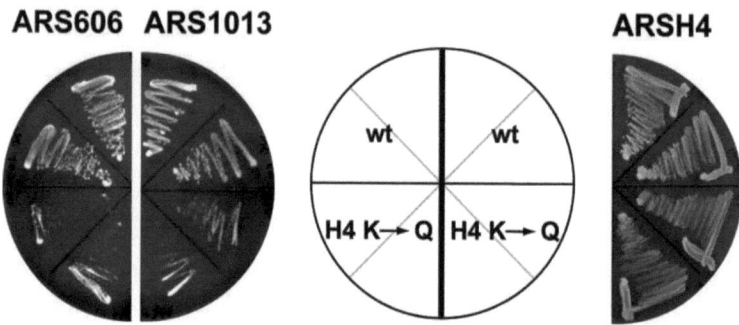

Fig. 29 Mutation of lysine to glutamine in the N-terminal tail of histone H4 caused a similar initiation defect as *sum1Δ* and *hst1Δ*
A wild-type (AEY3973) and a mutant strain (AEY3974) with K5, 8, 12 and 16 of histone H4 mutated to glutamine carrying *CEN4-URA3* plasmids with ARS606 (pAE1126), ARS1013 (pAE1081) or ARSH4 (pRS316) were streaked on minimal medium lacking uracil and incubated for three days at 30°C.

In summary, this suggested that increases in the acetylation level in *sum1Δ* and *hst1Δ* cells was responsible for the reduced origin activity of these ARS elements.

3.2.5 Histone H4 mutations did not cause synthetic lethality with *orc2-1*

The observation that histone deacetylation was responsible for full replication initiation of certain origins led to the question whether mimicking the deacetylation state by the histone H4 K5, 8, 12, 16 to glutamine mutation might cause synthetic lethality in combination with *orc2-1*, as has been seen with *sum1Δ*, *rfm1Δ* and *hst1Δ*. Therefore, *orc2-1* strains with *HHT1-HHF1* wild-type histone (AEY4202) or *HHT1-hhf1-10* (H4 K5, 8, 12, 16 to glutamine) histone mutant (AEY4239) *CEN4-TRP1* plasmids as the sole histone H3 and H4 source were constructed and tested for the growth ability on YM medium. The histone mutation did not cause a synthetic lethality in combination with *orc2-1*, but the strain showed a delayed growth compared to the *orc2-1* strain with wild-type histone H4 (Fig. 30). It can be concluded that while histone deacetylation was required for initiation function of certain origins and a deletion of the gene encoding the histone deacetylase Hst1 in *orc2-1* cells caused a synthetic lethality, a histone mutation that mimicked deacetylation of the acetylatable lysine residues of histone H4 was not sufficient to cause the same phenotype. Instead, other effects of *hst1Δ* beyond that of H4 K5, 8, 12, 16 deacetylation seemed to account for the synthetic lethality with *orc2-1*.

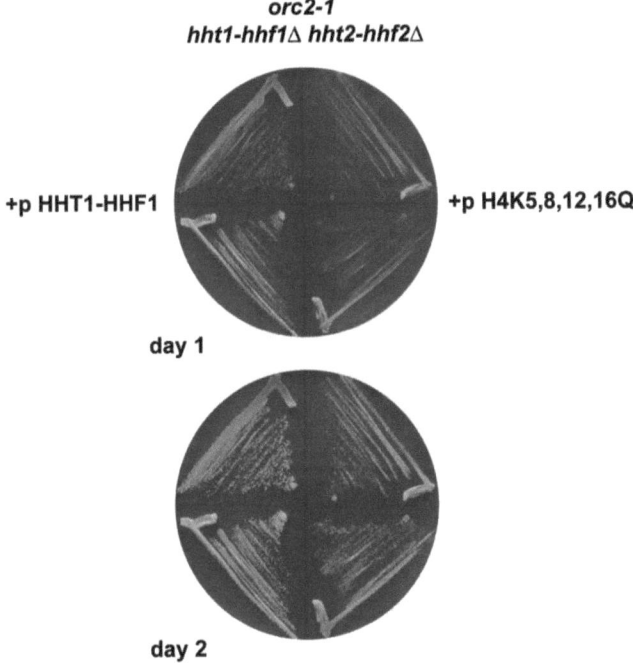

Fig. 30 Mutation of lysine to glutamine in the N-terminal tail of histone H4 caused a minor growth retardation in *orc2-1* cells
orc2-1 mutant strains (AEY4202) with either wild-type (pAE1193) or mutated histone H4 (lysine 5, 8, 12, 16 to glutamine) (pAE1192) *CEN-TRP1* plasmids as the sole histone H4 source were used for phenotypic analysis. The growth ability was documented for two days after streaking cells on minimal medium lacking tryptophan and incubation at 23°C.

3.3 sum1Δ caused synthetic growth defects in combination with mutations in sister chromatid cohesion factors

The results presented so far showed that Sum1, as part of the Sum1/Rfm1/Hst1 histone deacetylase complex, played a role in regulating replication initiation. Furthermore, in combination with *orc2-1*, *sum1*Δ, *rfm1*Δ and *hst1*Δ caused synthetic lethality. However, bioinformatics analysis (performed by T. Manke) indicated a binding of Sum1 in the proximity of ORC binding sites only for a subset of origins (Fig. 25), and not all of these seemed to be influenced to the same extent by Sum1. Because of these findings it could be assumed that the change in initiation function of these origins alone was not sufficient to explain the synthetic lethality between ORC and *sum1*Δ and thus, that ORC and Sum1 had similar functions in another cellular process.

Previous studies have proposed a function for ORC beyond replication initiation in regulating sister chromatid cohesion (Shimada & Gasser, 2007; Suter *et al.*, 2004). Therefore, it was interesting to determine whether Sum1 also contributed to sister chromatid cohesion. To address this, *SUM1* was deleted in different strains carrying mutations in cohesion genes. First, the effect of the absence of Sum1 was tested in a *S. cerevisiae* strain with the thermosensitive cohesin-subunit allele *smc3-42* (Klein *et al.*, 1999) that carried a *CEN4-URA3* plasmid containing *SMC3* (AEY4087). This was done by comparing an *smc3-42 SUM1* strain to an *smc3-42 sum1*Δ double mutant (AEY4128) for their ability to lose the *URA3*-marked *SMC3* plasmid on 5-FOA counterselective medium. While both strains showed reduced growth on 5-FOA medium, plasmid maintenance was stronger in the double mutant (Fig. 31A). However, either strain was able to lose the *SMC3* plasmid, and therefore it could be concluded that *sum1*Δ did not cause synthetic lethality with *smc3-42*.

Next, it was determined whether *sum1*Δ had an influence on the maximal permissive growth temperature of the *smc3-42* mutation. Serial dilutions of wild-type (AEY2), *smc3-42* (1341) and *sum1*Δ (AEY3358) single mutants or *smc3-42 sum1*Δ double mutant cells were tested for a temperature phenotype at 23 °C, 27 °C, 30 °C, 34 °C, 37 °C on full medium. No difference at all temperatures was observed between wild-type and *sum1*Δ single mutant strains. As expected, growth of the *smc3-42* was severely impaired at 34 °C and completely abolished at 37 °C. An additional *SUM1* deletion slightly enhanced in the thermosensitive phenotpye of *smc3-42* as shown by a growth decrease at 30 °C while there was no difference at lower temperatures (Fig. 31B).

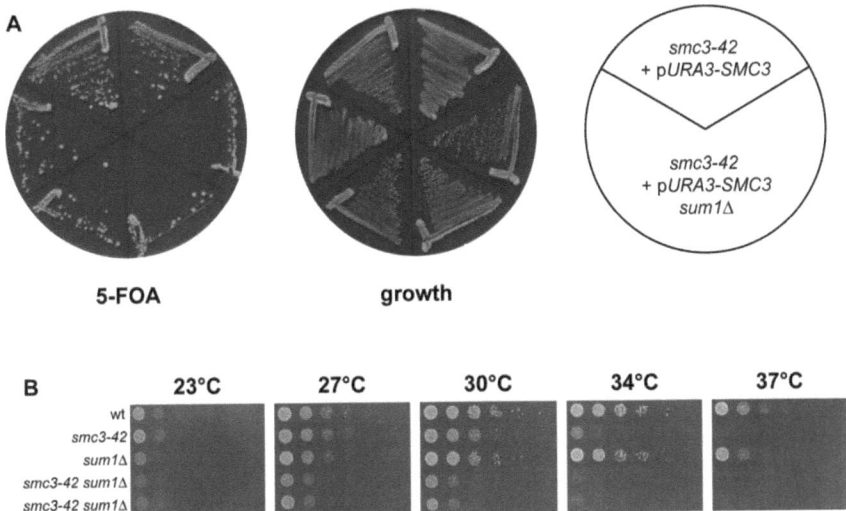

Fig. 31 *sum1Δ* showed a slight synthetic growth defect with the sister chromatid cohesion mutant *smc3-42*
The influence of a *SUM1* deletion on the growth ability of a cohesion mutant was analysed. (A) *smc3-42 SUM1* (AEY4087) and *smc3-42 sum1Δ* (AEY4128) *S. cerevisiae* cells carrying p*URA3-SMC3* (pAE568) were tested on 5-FOA medium for their ability to lose the *SMC3* plasmid. Strains were incubated for three days at 23°C. (B) Serial dilutions of a wild-type (AEY2) strain, *smc3-42* (AEY1341) and *sum1Δ* (AEY3358) single or *smc3-42 sum1Δ* double mutant (AEY4128) strains were grown on full medium and incubated for two days at 23°C, 27°C, 30°C, 34°C and 37°C and tested for temperature sensitivity.

Further, a second approach was used to examine a possible role of Sum1 in the mechanism of sister chromatid cohesion. Instead of a mutant allele, the non-lethal deletion of *CTF18*, part of the alternative replication factor C (RFC) complex that is required for sister chromatid cohesion (Mayer *et al.*, 2001), was combined with *sum1Δ*. *ctf18Δ* was chosen, because previously a strong genetic interaction between the Ctf18-RFC complex and *orc2-1* has been shown (Suter *et al.*, 2004). If Sum1 interacted with Ctf18 in sister chromatid cohesion, a *sum1Δ ctf18Δ* double deletion would be expected to affect this mechanism and cause a severe growth defect. To address this, growth of serial dilutions of wild-type (AEY2, AEY3), *sum1Δ* (AEY3358) and *ctf18Δ* (AEY4199) cells were compared to different *sum1Δ ctf18Δ* double mutants (AEY4206-4208) that originated from a genetic cross of the parent strains. As one of the parent strains had an ADE^+ and the other an ade^- phenotype and the latter is known to cause slightly delayed growth, ADE^+ and ade^- double mutants were used in this experiment in order to rule out effects due to this difference. While all three tested *sum1Δ ctf18Δ* strains seemed to display reduced growth after one day of incubation at 30 °C, this observation could not be verified after two days (Fig. 32).

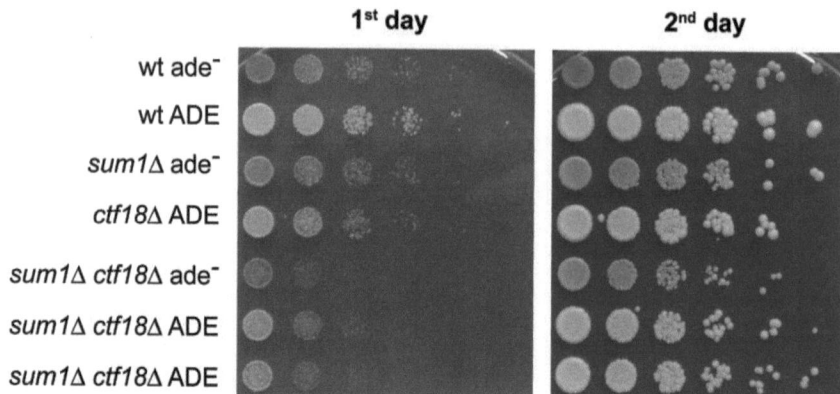

Fig. 32 sum1Δ showed a slight growth defect with ctf18Δ
Serial dilutions of wild-type (AEY2, AEY3) strains, sum1Δ (AEY3358) and ctf18Δ (AEY4199) single or sum1Δ ctf18Δ double mutant (AEY4206-4208) strains were grown on full medium and incubated at 30°C. Cell growth was documented after one or two days, respectively. To rule out adenine auxo- or prototrophic effects on cell growth ADE⁺ and ade⁻ strains were used for wild-type controls and double mutants.

Taken together, sum1Δ caused a mild synthetic growth defect in combination with smc3-42 and a minor defect in ctf18Δ. This suggested a minor role for Sum1 in sister chromatid cohesion.

3.4 Factors that bind in proximity to ORC binding sites were essential for viability of *orc2-1* mutant strains

The results presented so far have shown that regulation of silencing function at *HML* as well as replication origin activity do not solely depend on the origin recognition complex but also on auxiliary factors such as Sum1, which bind in the proximity to ORC to origin sequences. Notably, *HML* is a single chromosomal locus that allows binding of a limited number of proteins besides ORC. In contrast to that, there are multiple replication origins within the *S. cerevisiae* genome, of which some are more efficient than others. However, while Sum1 was shown to influence *HML-SS* ΔI silencing (see. 3.1.3), only a subset of origins displayed a dependence on Sum1 (see 3.2.2). Therefore the question arose, which other factors bind in the vicinity of several other ORC binding sites, and it was interesting to determine whether these proteins facilitate ORC's function in silencing or replication initiation.

3.4.1 *dat1Δ*, *gat3Δ* and *rgm1Δ* were synthetically lethal with an *orc2-1* mutation

Bioinformatics analyses revealed other DNA binding proteins that share binding to the same intergenic regions with ORC (T. Manke, personal communication 2005) and therefore could probably influence origin activity. It was calculated whether binding of the respective factor and ORC in proximity to each other, which in the following will also be called "co-occurrence", was non-coincidental (all bioinformatics data were kindly provided by T. Manke from the Max Planck Institute for Molecular Genetics, Berlin). Depending on the assessed p-value threshold (p = 0.001 or p = 0.01) for binding of ORC and the second factor to the DNA, two or three factors were identified that bind non-coincidental to the same intergenic regions. The most significant co-occurrence (p-value = 6.59×10^{-9}) with ORC was shown for the GATA-family zinc-finger-containing protein Gat3 (Cox *et al.*, 1999). While ChIP data (Harbison *et al.*, 2004; Wyrick *et al.*, 2001) identified 305 targets that are bound by ORC and 136 by Gat3, 27 of these intergenic regions were in common, which represents a statistically relevant overlap. The factor with the next best consistent co-occurrence (p-value = 3.39×10^{-5}) was the DNA-binding protein Dat1 (Winter & Varshavsky, 1989). The intergenic regions with the most probable non-coincidental binding of ORC and Dat1 or Gat3 are listed in Table 13 and Table 14 and in the p-value plots for Dat1 (Fig. 33) and Gat3 (Fig. 34). Low values on the x- and y-axes indicated a strong potential for a non-coincidental binding of both factors to the same region.

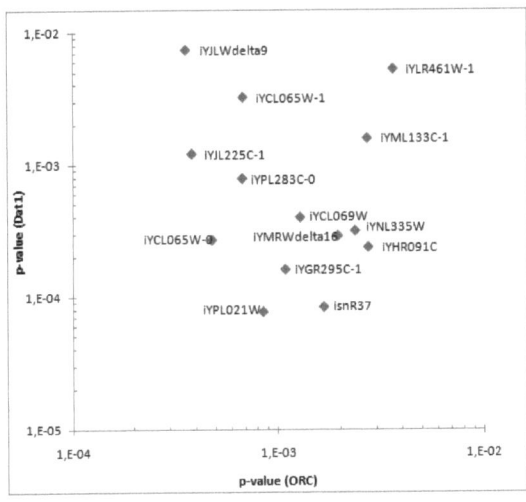

Fig. 33 Search for intergenic regions that bind both Dat1 and ORC
Plot of p-values for Dat1 binding, $P < 0.001$ (Harbison *et al.*, 2004) versus ORC binding, $P < 0.05$ (Wyrick *et al.*, 2001).

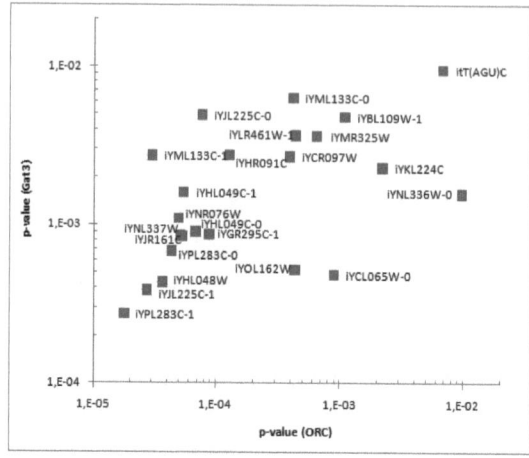

Fig. 34 Search for intergenic regions that bind both Gat3 and ORC
Plot of p-values for Gat3 binding, $P < 0.001$ (Harbison et al., 2004)versus ORC binding, $P < 0.05$ (Wyrick et al., 2001).

When the p-value threshold was changed from $p = 0.001$ to $p = 0.01$, the p-values for co-occurrence with ORC binding also changed, for Gat3 to 1.79×10^{-9} and Dat1 to 7.16×10^{-6}. Then the second best candidate after Gat3 was the putative transcriptional repressor Rgm1 (Estruch, 1991) (p-value = 1.52×10^{-8}).

Table 13: Intergenic regions with high co-occurrence of Dat1 and ORC binding

Intergenic region	Dat1 p-value	ORC p-value
iYPL021W	7.76×10^{-5}	8.57×10^{-4}
iYCL065W-0	2.72×10^{-4}	4.85×10^{-4}
isnR37	8.41×10^{-5}	1.68×10^{-3}
iYGR295C-1	1.63×10^{-4}	1.10×10^{-3}
iYJL225C-1	1.22×10^{-3}	3.86×10^{-4}
iYCL069W	4.03×10^{-4}	1.30×10^{-3}
iYPL283C-0	7.92×10^{-4}	6.81×10^{-4}
iYMRWdelta	2.91×10^{-4}	1.97×10^{-3}
iYHR091C	2.39×10^{-4}	2.75×10^{-3}
iYNL335W	3.18×10^{-4}	2.38×10^{-3}
iYCL065W-1	3.26×10^{-3}	6.87×10^{-4}
iYJLWdelta9	7.43×10^{-3}	3.61×10^{-4}
iYML133C-1	1.60×10^{-3}	2.73×10^{-3}
iYLR461W-1	5.30×10^{-3}	3.65×10^{-3}

Table 14: Intergenic regions with high co-occurrence of Gat3 and ORC binding

Intergenic region	Gat3 p-value	ORC p-value
iYPL283C-1	1.78×10^{-5}	2.73×10^{-4}
iYJL225C-1	2.73×10^{-5}	3.86×10^{-4}
iYHL048W	3.65×10^{-5}	4.32×10^{-4}
iYPL283C-0	4.36×10^{-5}	6.81×10^{-4}
iYJR161C	5.12×10^{-5}	8.49×10^{-4}
iYNL337W	5.32×10^{-5}	8.42×10^{-4}
iYGR295C-1	4.94×10^{-5}	1.10×10^{-3}
iYHL049C-0	6.78×10^{-5}	8.98×10^{-4}
iYNR076W	8.79×10^{-5}	8.64×10^{-4}
iYML133C-1	2.99×10^{-5}	2.73×10^{-3}
iYHL049C-1	5.43×10^{-5}	1.60×10^{-3}
iYOL162W	4.35×10^{-4}	5.23×10^{-4}
iYHR091C	1.28×10^{-4}	2.75×10^{-3}
iYJL225C-0	7.66×10^{-5}	4.91×10^{-3}
iYCL065W-0	8.99×10^{-4}	4.85×10^{-4}
iYCR097W	3.91×10^{-4}	2.69×10^{-3}
iYLR461W-1	4.39×10^{-4}	3.65×10^{-3}
iYMR325W	6.47×10^{-4}	3.62×10^{-3}
iYML133C-0	4.19×10^{-4}	6.32×10^{-3}
iYKL224C	2.18×10^{-3}	2.29×10^{-3}
iYBL109W-1	1.09×10^{-3}	4.80×10^{-3}
iYNL336W-0	9.70×10^{-3}	1.57×10^{-3}
itT(AGU)C	6.74×10^{-3}	9.53×10^{-3}

To determine whether Dat1, Gat3 and Rgm1 affected ORC function *in vivo*, deletion mutants of all three factors were combined using genetic crosses with the temperature-sensitive *orc2-1* allele and phenotypically analysed. Interestingly, only if previously provided with an *ORC2*-carrying plasmid with a different selection marker but not a vector control, *orc2-1 dat1Δ* (AEY3971 / AEY3972), *orc2-1 gat3Δ* (AEY3968 / AEY3970) and *orc2-1 rgm1Δ* (AEY4077 / AEY4078) strains were able to lose an *URA3*-marked *ORC2* plasmid on counterselective medium (5-FOA). (Fig. 35). This demonstrated a synthetic lethality of *orc2-1* with either gene deletion suggesting that they played a role in replication initiation.

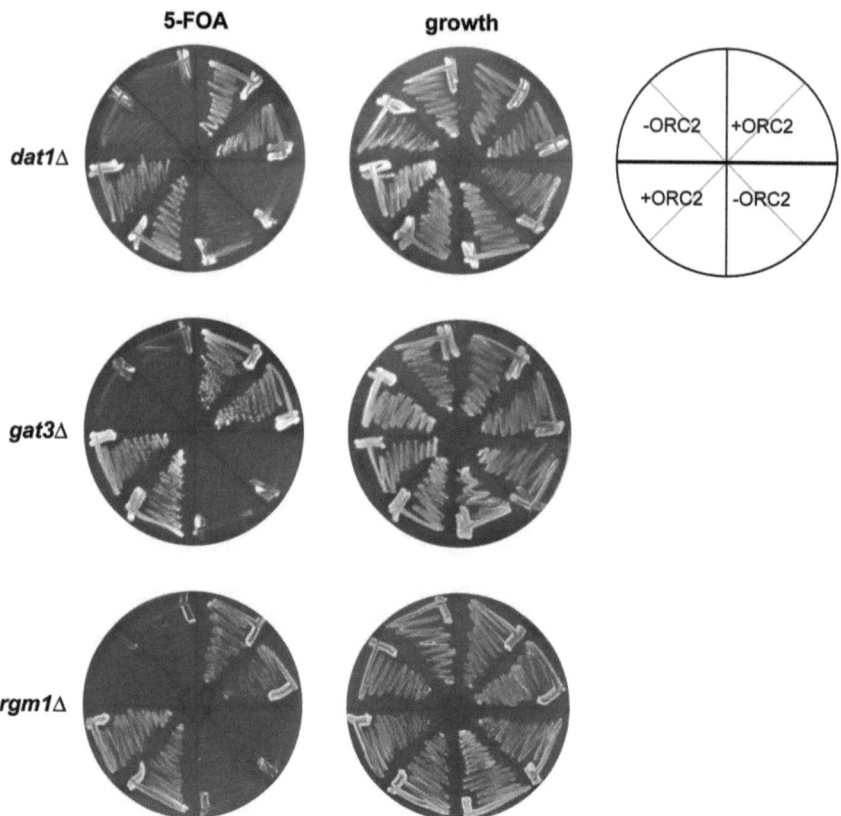

Fig. 35 *dat1Δ*, *gat3Δ* and *rgm1Δ* showed synthetic lethality with *orc2-1*
orc2-1 dat1Δ (AEY3971 / AEY3972), *orc2-1 gat3Δ* (AEY3968 / AEY3970) and *orc2-1 rgm1Δ* (AEY4077 / AEY4078) strains carrying p*URA3-ORC2* (pAE1316) were tested on 5-FOA medium for their ability to lose the *ORC2* plasmid when additionally provided with a *LEU2*-marked *ORC2* plasmid (pAE1315) or empty *LEU2* vector (pRS315). Strains were incubated for three days at 23°C.

3.4.2 *dat1Δ* and *gat3Δ* did not influence origin activity on chromosome X

Since several of the intergenic regions with high rankings for ORC and Dat1 or Gat3 binding constitute ARS sequences on chromosome X (Table 15), it was interesting to determine whether Dat1 or Gat3 influenced initiation function of these origins.

Table 15: Intergenic regions on chromosome X with high co-occurrences

Intergenic region	Corresponding ARS	Proposed co-occurrences
iYJL225C-1	ARS1001	ORC-Dat1, ORC-Gat3
isnR37	ARS1009	ORC-Dat1
iYJLWdelta9	ARS1014	ORC-Dat1
iYJR161C	ARS1025	ORC-Gat3

Plasmid maintenance of *dat1Δ* and *gat3Δ* strains with *CEN4-URA3* plasmids containing ARS1001, ARS1009, ARS1014, or ARS1025 as the sole origin were compared to that of the corresponding wild-type strain (AEY3). If Dat1 or Gat3 had a function as auxiliary factors for replication initiation at these origins, plasmid maintenance would have been expected to be impaired in the deletion strains. However, neither the growth test of serial dilutions (exemplified in Fig. 36A) nor the determination of plasmid loss (Fig. 36B) revealed any significant differences between the *dat1Δ* and *gat3Δ* strains and the wild-type control. This suggested that Dat1 and Gat3 did not influence origin activity on chromosome X.

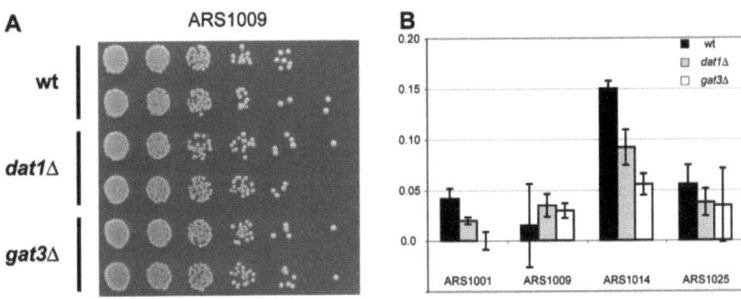

Fig. 36 Dat1 and Gat3 were dispensable for ARS activity of origins from chromosome X
Plasmid maintenance of *CEN4-URA3* plasmids with ARS1001 (pAE1276), ARS1009 (pAE1278), ARS1014 (pAE1279) and ARS1024 (pAE1280) as their sole origin was analysed in wild-type (AEY3), *dat1Δ* (AEY3915) and *gat3Δ* (AEY3917) strains. (A) Example of a serial dilution experiment with strains carrying an ARS1009 URA3 plasmid. Cells were incubated on minimal medium lacking uracil for two days at 30°C. (B) Measurement of plasmid loss rates. Error bars represent the standard deviations of three independent determinations.

3.4.3 *dat1Δ* and *gat3Δ* did not influence silencing of the telomeres or the *HML* locus

Having demonstrated that *dat1Δ* and *gat3Δ* caused synthetic lethality in *orc2-1* cells on the one hand, but that replication function on chromosome X was not impaired on the other hand, it was still to be determined what might have caused the lethality. A more detailed analysis of the intergenic regions that showed a high co-occurrence for ORC and Dat1 or Gat3 binding revealed that many of those regions were located at or near the telomeres. It was known that *orc2-1* cells are defective in telomeric silencing (Fox *et al.*, 1997). This led to the assumption that Dat1 and Gat3 could also contribute to telomeric silencing. Therefore, synthetic telomeric *URA3* or *ADE2-URA3* constructs were introduced into *dat1*Δ (AEY4121 / AEY4124) and *gat3*Δ (AEY4122 / AEY4125) strains. No differences in the growth on counterselective medium (5-FOA) or minimal medium lacking uracil of the deletion strains compared to a wild-type control (AEY4120) with the TEL VII L*::URA3* construct was observed (Fig. 37).

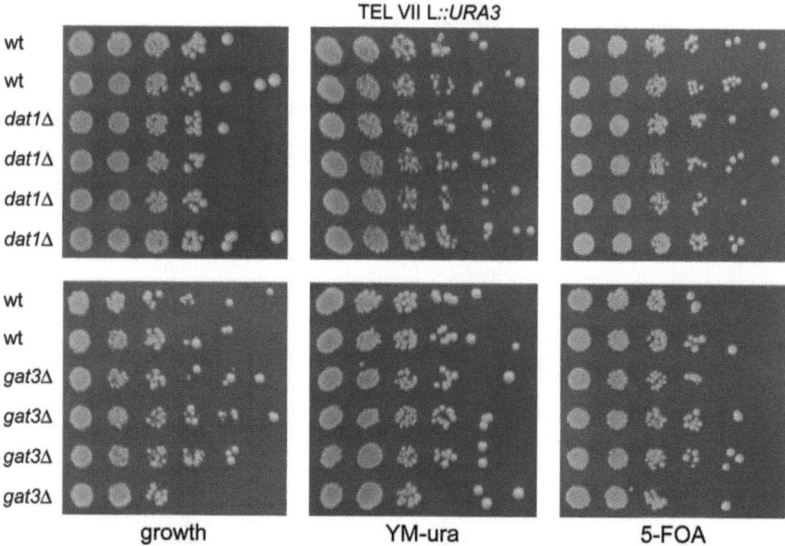

Fig. 37 *dat1*Δ **and** *gat3*Δ **did not affect telomeric** *URA3* **silencing**
Serial dilutions of wild-type (AEY4120), *dat1*Δ (AEY4121) and *gat3*Δ (AEY4122) strains containing a synthetic TEL VII L*::URA3* construct were tested on 5-FOA medium for their ability to silence the telomeric *URA3* gene. Cells were grown for two days at 30°C.

Telomeric *ADE2* silencing was determined by monitoring the colony colour of the deletion strains compared to the wild-type control (AEY4123). Loss of silencing of the TEL VII L*::ADE2-URA3* construct resulted in a shift from red to white colonies. Therefore, an increase in white or sectored colonies due to partial silencing was a sign for impaired telomeric silencing. The percentage of red, sectored and white colonies was calculated and the results were illustrated in a bar graph (Fig. 38). While the fraction of white colonies was similar in all three strains, there was a slight decrease in red and an increase in sectored colonies in *dat1*Δ and *gat3*Δ cells. This indicated a minor reduction in telomeric *ADE2* silencing in the deletion strains. Because this minor difference in telomeric silencing was not seen with the synthetic TEL VII L::*URA3* construct, it was next investigated whether natural telomeric *URA3* silencing was influenced by Dat1 or Gat3.

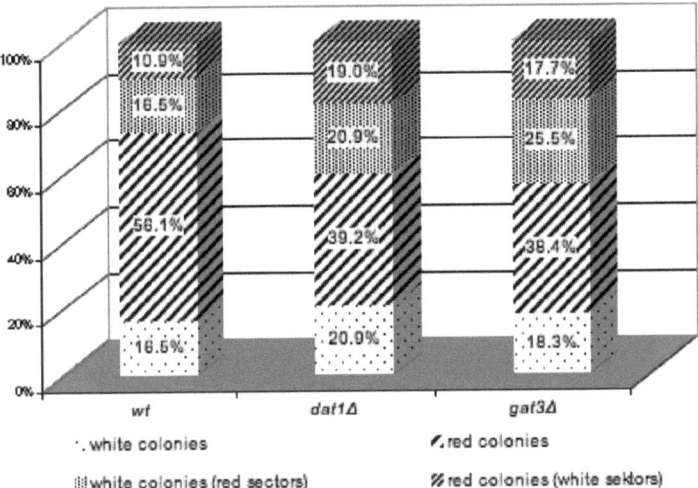

Fig. 38 *dat1Δ* and *gat3Δ* caused a minor reduction in telomeric *ADE2* silencing
Wild-type (AEY4123), *dat1Δ* (AEY4124) and *gat3Δ* (AEY4125) strains containing a synthetic TEL VII L::*ADE2-URA3* construct were tested for their ability to silence the telomeric *ADE2* gene by determining amounts of red, white and sectored colonies. Cells were grown on full medium for two days at 30°C.

Therefore, serial dilutions of wild-type (AEY2156 / 2159), *dat1Δ* (AEY4156 / AEY4157) and *gat3Δ* (AEY 4158 / 4159) strains containing TEL IX L::*URA3* / TEL XI L::*URA3* strains were analysed for their growth ability on counterselective medium (5-FOA) or minimal medium lacking uracil. Again, no difference was observed (Fig. 39). Taking all three experimental approaches together, *dat1Δ* and *gat3Δ* did not affect telomeric silencing.

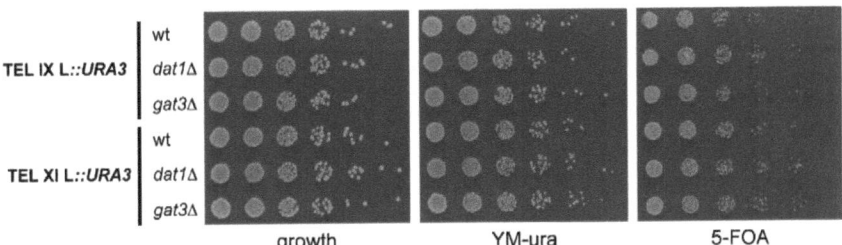

Fig. 39 *dat1Δ* and *gat3Δ* did not affect silencing of natural telomeric *URA3* insertions
Serial dilutions of wild-type (AEY2156 / AEY2159), *dat1Δ* (AEY4156 / AEY4157) and *gat3Δ* (AEY4158 / AEY4159) strains containing a natural TEL IX L::*URA3* / TEL XI L::*URA3* construct were tested on 5-FOA medium for their ability to silence the telomeric *URA3* gene. Cells were grown for two days at 30°C.

Besides intergenic regions located at or near telomeres, the search for co-occurrence of ORC and Dat1 / Gat3 binding also revealed regions near the *HM* loci. In order to test whether Dat1 or Gat3 influenced *HM* silencing, genetic crosses were performed to combine deletions of these factors with mutations in the *HML* region. *MAT*α *dat1*Δ (AEY3915) or *MAT*α *gat3*Δ (AEY3917) cells were crossed with a *MAT***a** *HML* ΔI (AEY3388) or *MAT***a** *HML acs⁻* ΔD ΔI (AEY3398) strains and *MAT***a** cells with the respective gene deletion and wild-type or mutant *HML*-E silencer were obtained. If Dat1 or Gat3 affected *HML* silencing, strains with a sensitized *HML*-E silencer (*HML* ΔI or *HML acs⁻* ΔD ΔI) should have shown reduced mating ability compared to those with wild-type *HML*-E. However, with neither deletion mutant a difference was observed (data not shown), indicating that Dat1 and Gat3 were not involved in *HM* silencing.

To conclude, although *dat1*Δ, *gat3*Δ and *rgm1*Δ caused synthetic lethality with *orc2-1*, it was not possible to find a connection of these three factors in any observed cellular mechanism with the origin recognition complex. Neither origin function nor telomeric or *HML* silencing seemed to be influenced by Dat1, Gat3 and Rgm1. Therefore, it remains unclear what caused the lethality of the double mutants, while single gene deletion and *orc2-1* strains are viable. However, as the binding sites for Dat1 and Gat3 are AT-rich and similar nucleotide enrichment is found at the telomeres, it was possible that high co-occurrence values might be coincidental due to that fact. Therefore, the bioinformatics analysis searching for factors that bind in the vicinity of ORC was repeated excluding telomeric regions (calculations again performed by T. Manke). Depending on the calculation method for the highest similarity with respect to ORC binding (p-value with threshold or correlation without threshold) a new list of potentially interacting factors evolved (Table 16).

Table 16: List of factors with high probability of binding proximal to ORC

rank	factor	p-value with threshold (0.001)	Factor	correlation without threshold
1	ORC	0	ORC	1.0
2	Dat1	1.7×10^{-6}	Sum1	0.26
3	Spt23	2.2×10^{-5}	Pdc2	0.18
4	Ume1	7.6×10^{-4}	Mig3	0.18
5	Ime4	2.6×10^{-4}	Dig1(alpha)	0.18
6	Gzf3	1.0×10^{-3}	Arr1	0.16
7	Yox1	1.1×10^{-3}	Hir2	0.16
8	Gal4	1.1×10^{-3}	Yox1	0.16
9	Hir2	1.7×10^{-3}	Rlm1	0.15

Interestingly, without including telomeres in the threshold analysis, Gat3 lost eight of ten co-occurrences, thus changing the calculated p-value dramatically from 3×10^{-11} with telomeres to 1×10^{-2} without (T. Manke, personal communication 2007). The number of ORC and Dat1 co-occurrences is reduced from six to five, leaving a significant p-value of 1.7×10^{-6} (with telomeres 2.9×10^{-7}). According to this determination method, Dat1 was the factor that most probably binds in the vicinity of ORC, while there was no indication for Gat3 or Rgm1 co-occurrence with ORC.

4. Discussion

The focus of this thesis was to investigate the contribution of the origin recognition complex (ORC) and the transcriptional repressor Sum1 to the two distinct processes of replication initiation and silencing of the silent mating type locus *HML* in *Saccharomyces cerevisiae*. Common features such as the involvement of auxiliary factors that bind in the vicinity of ORC as well as the influence of histone deacetylation complexes were included in the analyses. It was found, that binding sites for ORC, Sum1 and the DNA-binding factor Rap1 within a synthetic, minimal *HML*-E silencer were required for functional *HML* silencing. Mutations in the consensus sequences or *in trans* mutations of the respective proteins caused a severe loss of silencing function.

Furthermore, Sum1 was required for replication initiation at a subset of origins bound by Sum1 proximal to ORC. It was shown that histone H4 acetylation at these origins was influenced by Sum1 and Hst1, which are part of a HDAC complex comprising the bridging factor Rfm1. Deletions of all three components of this HDAC caused a synthetic lethal phenotype in combination with the *orc2-1* mutation. Similar observations were made for other factors that are suggested to be co-localized with ORC at several intergenic regions. However, as these factors apparently were not involved in sister-chromatid cohesion, telomeric or *HM* silencing, the reason for the observed synthetic lethality remains unknown.

4.1 A synthetic *HML*-E construct was sufficient for *HML* silencing

Repression of mating-type information at the silent mating type loci *HMR* and *HML* in *S. cerevisiae* is necessary to maintain cell-type identity in haploid strains. The mating-type is determined by alternative alleles of the *MAT* locus, but additional copies of the **a** and α genes are located at the silent mating-type loci *HML* and *HMR*. Impaired silencing of these loci results in pseudo-diploid yeast cells expressing **a**/α information, thus impairing their mating with cells of the opposing mating type. *HM* silencing is buffered towards mutations *in cis* and *in trans* that affect the *HM* silencers. Here, a synthetic, minimal *HML*-E silencer was generated that lacks the functional redundancy of natural *HML*-E. It will therefore be useful for studies to identify new factors that are involved in the regulation of *HML* silencing, but so far escaped identification due to the redundancy of the wild-type *HML* silencer.

The construction of a synthetic *HML*-E silencer conducted here was inspired by a classical study in which a minimal silencer for the other silent mating-type locus, *HMR*, was generated

(McNally & Rine, 1991). Similarly to that silencer, the minimal *HML*-E version presented here consisting of a Rap1 binding site, an ACS and the D2 element alone showed a silencing ability comparable to that of wt *HML*-E, thus establishing for the first time that these three domains alone are sufficient for *HML* silencing. A quantitative analysis revealed, that while the synthetic wild-type *HML*-E version exhibited 100 % silencing, scrambling of the nucleotide composition without affecting the binding sites for Rap1, ORC and Sum1 (*HML-SS* ΔI) was sufficient to reduce silencing to 60% of wild-type function. This indicates that the core *HML*-E silencer contains so far unknown elements, which make up 40 % of wild-type silencing and that were abrogated in *HML-SS* ΔI.

Nevertheless, a substantial difference in silencing function between the previously generated synthetic *HMR*-E silencer (McNally & Rine, 1991) and the new minimal *HML*-E silencer was observed. While the former resulted in approximately 15 % of wild-type *HMR*-E silencing, the synthetic *HML*-E silencer exhibited 60 % of wild-type function. This difference may be explained by the extent of the deleted sequences surrounding the E silencers. In case of the minimal *HMR*-E silencer, 800 bp were replaced by the synthetic construct, whereas the synthetic *HML*-E silencer replaced a much shorter, 223 bp sequence. At *HML*-E, it was not possible to truncate the synthetic *HML*-E silencer to the same extent as it was done for *HMR*-E without affecting the neighbouring gene *VBA3* and the W region of *HML*. If regions directly adjacent to the core E-silencer contributed to silencing function of *HMR*-E, this influence was abrogated at the synthetic *HMR*-E. While it is not likely that the *VBA3* gene upstream of *HML*-E contains silencing elements, it cannot be completely excluded that the W region, which is also present at the *MAT* locus but not at *HMR*, contributes to *HML* silencing. An influence of this element would still be present at the synthetic *HML*-E silencer. This could explain why the minimal *HML*-E silencer had stronger silencing function than the minimal *HMR*-E silencer.

Furthermore, the *HML-SS* ΔI silencer was sensitive to mutations in any one of the three silencer elements, thus also showing that they were necessary for silencing, and that the functional redundancy of natural *HML*-E has been eliminated in this construct. However, some unexpected observations concerning the *trans* requirements for silencing of *HML-SS* ΔI were made.

As expected, the silencing-defective *rap1-12* allele (Sussel & Shore, 1991) caused as strong derepression as a mutation of the Rap1 binding site of *HML-SS* ΔI, which was consistent with its known role in *HML* silencing. The strong effect of *rap1-12* may also be related to the fact that there is an additional Rap1 binding site in the UAS of the α2 gene, which has been shown to serve as a proto-silencer in *HML* silencing (Boscheron *et al.*, 1996).

Surprisingly, while mutation of the putative ORC binding site (ACS) caused strong *HML* derepression, two mutant *orc* alleles, *orc2-1* (Foss *et al.*, 1993) and *orc5-1* (Loo *et al.*, 1995a), only caused a mild loss of silencing. These alleles were originally isolated based on their ability to cause derepression of a version of natural *HMR*-E, and they also derepress synthetic *HMR*-E (Foss *et al.*, 1993; Fox *et al.*, 1995). One could therefore argue that the *HML*-E ACS for some reason is not sensitive to these particular *orc* alleles. In fact, a recent genome-wide study of ORC binding showed that not every chromosomal origin is equally sensitive to *orc2-1* (Shor *et al.*, 2009). However, the ACS of synthetic *HML*-E is identical to that of synthetic *HMR*-E. Also, in a highly sensitive silencing assay, natural *HML*α, which also contains the same ACS, showed slight derepression by *orc2-1* as measured by the α-mating ability of a strain lacking coding information at *MAT* (Loo *et al.*, 1995a). Therefore, this suggests that the sequences surrounding the ACS at the *HML*-E silencer determine whether it is sensitive to the *orc* alleles or not. It is also possible that "non-silencer replicator origins" remain in the synthetic silencer, as has been described for the natural *HMR*-E silencer (Palacios DeBeer & Fox, 1999) and despite the efforts to remove them in the synthetic *HML*-E construct. In light of this, there may exist a competition between the silencer ACS at *HML*-E and other putative ORC binding sites in the vicinity, which may be responsible for the unexpected insensitivity of silencing to *orc2-1*. Of note, this would have to be a competition between silencer and non-silencer ORC binding sites at *HML*, rather than between silencer and non-silencer origins at *HMR*, because *HML*-E is not a chromosomal replication origin, but is passively replicated by a replication fork originating from a nearby origin, ARS305 (Vujcic *et al.*, 1999).

Alternatively, in light of a recent study showing that ORC binding spread throughout the *HMR* silent domain rather than being restricted to the *HMR* silencers (Ozaydin & Rine, 2009), it is also possible that ORC similarly binds *HML* beyond the silencer, and that this binding, and thus the contribution of ORC to silencing, is not abrogated by *orc2-1* and *orc5-1*.

Furthermore, it was observed that the mutation of the D2 site of *HML-SS* ΔI caused strong derepression, but that the absence of Sum1, which has previously been shown to bind to D2, caused only a minor amount of derepression (Irlbacher *et al.*, 2005). Earlier genetic evidence for the involvement of Sum1 in *HML* silencing demonstrated that it caused derepression of natural *HML*-E that was sensitized by the deletion of the Rap1 or ACS elements, but not the D element. Thus, the difference in sensitivity to Sum1 between natural and synthetic *HML*-E may lie in the sequence differences between the two silencers. It is also possible that the D2 element binds another protein in addition to Sum1, and that both need to be mutated to cause strong *HML* derepression. Further work will be required to identify such a factor.

The contribution of ORC and Sum1 to *HML-SS* ΔI was also confirmed in a *mat*Δ strain background. As expected, *HML-SS* ΔI seemed to be derepressed upon deletion of *SUM1*, indicated by reduced mating ability on a *MAT*α lawn. An even more severe effect was observed for an *orc2-1* mutation in a *mat*Δ *HML-SS* ΔI strain. Deletion of the *MAT*α1 promoter and *HML* derepression enabled the strains to mate with a *MAT*a strain. The deletion of *SUM1* enhanced this α mating ability suggesting an increase of α mating type information due to reduced *HML-SS* ΔI silencing. Surprisingly, an *orc2-1* mutation did not enhance α mating ability like *sum1*Δ did. In contrast, a *mat*Δ *HML-SS* ΔI *orc2-1* strain seemed to be completely unable to mate with a *MAT*a strain, indicating that in this context wild-type ORC is required for the expression of α mating type information from *HML-SS* ΔI.

4.2 *HML-SS* ΔI was sensitive for some, but not all previously identified mutations that influence *HML* silencing

A number of proteins involved in *HM* silencing have been identified over the years. Among these, one can distinguish between those generally essential for silencing, like the Sir2, Sir3 and Sir4 proteins (Rine & Herskowitz, 1987), those that have an important function in silencing like Sir1 (Pillus & Rine, 1989; Rine & Herskowitz, 1987), and factors whose contribution to silencing is only apparent upon mutation or deletion of a second factor (reviewed in (Rusche et al., 2003)). The *HML-SS* ΔI silencer developed here is a minimal silencer that provides a sensitized background to identify novel regulators of *HML* silencing. Notably, as for synthetic *HMR*-E, the novel silencer presented in this work was fully sensitive to the deletion of *SIR1*. In addition, *dot1*Δ, which derepresses natural *HML* only in a *sir1*Δ background (van Welsem et al., 2008), caused complete derepression of *HML-SS* ΔI, thus providing a first example for a factor whose effect only becomes apparent in the sensitized background. Mechanistically, this may be explained by a less robust binding of the SIR proteins to synthetic *HML*-E, such that they are more easily redistributed to euchromatic sites when genome-wide H3 K79 methylation is lost in the absence of Dot1.

Surprisingly, *HML-SS* ΔI was sensitive to some, but not other silencing factors. For instance, its silencing was refractory to *asf1*Δ and *sas2*Δ, although both cause derepression of natural

HML in *sir1Δ* cells (Meijsing & Ehrenhofer-Murray, 2001; Reifsnyder et al., 1996). This indicates that the sensitization by *HML-SS* ΔI is distinct from that of the absence of Sir1, and that synthetic *HML*-E opens up the possibility of identifying novel silencing factors whose effect has so far been masked by genetic redundancy. The simplicity of the synthetic silencer will thus facilitate new insights into the mechanisms of transcriptional silencing.

4.3 High-throughput genetic screens as useful methods to identify novel silencing factors?

In order to find novel factors for *HML* silencing, an adapted genetic screen (Tong et al., 2001) with the *S. cerevisiae* deletion library, which comprises ~ 5000 strains, was used. After a multi-step selection procedure, each of the gene deletions was combined with the minimal *HML*-E silencer and the influence of the mutation on *HML* silencing by the synthetic silencer was measured. This was achieved by analyzing the mating ability of *MAT***a** *HML-SS* ΔI strains carrying the individual gene deletions. However, experiments revealed that this was not a trivial task. During the project presented here, the *HML-SS* ΔI allele was introduced into two different strains commonly used for genetic crosses with the deletion library. The first strain failed to permit an adequate genetic background for the screening experiment as described in the results section.

Therefore, a second strain with different and more selection markers was used in a repetition of the screen. Using this second strategy the background "noise" of diploid and *MAT*α cells, which were obtained with the first strain and obstructed mating analysis, was significantly reduced. However, the setup of the screen, which was modified to analyse changes in *HML* silencing, resulted in another problem. In principle, the *can1Δ::STE2pr-his5* construct used here allows to select for *MAT***a** cells, since the *STE2* α-factor receptor is an **a**-specific gene and thus is only expressed in *MAT***a** cells. As the *Schizosaccharomyces pombe his5* gene, which corresponds to the *S. cerevisiae HIS3* gene, is fused to the *STE2* promoter, *his5* is thus only expressed in *MAT***a** cells. Because the deletion library strains and the synthetic lethality screen strain, which were used for the genetic crosses in this screen, have a *his3* genotpye, only *MAT***a** cells with an active *STE2* promoter express *his5* from the *can1Δ::STE2pr-his5* construct and can grown on minimal medium lacking histidine. If a gene deletion from the deletion library does not affect silencing of the synthetic *HML*-E allele, the selection works as described. However, if a mutation reduces *HML-SS* ΔI silencing, *MAT***a** cells express **a** mating-type information from the *MAT* locus and, in addition, α information from the derepressed *HML* locus. Then, the **a**-specific *STE2* promoter is repressed by the **a**1/α2 repressor and these pseudo diploid **a**/α cells fail to grow on minimal medium unless supplemented with histidine. A genetic screen with medium lacking histdine to select for haploid *MAT***a** cells would therefore select against *MAT***a** *HML-SS* ΔI mutants that carry the deletion of a factor that influences *HML* silencing in the sensitized background.

For this reason, since the aim of the genetic screen was to identify precisely these factors, the selection procedure had to be adjusted in order not to forfeit identification of positive candidates that are involved in *HML* silencing. Consequently, this meant that it was only possible to select with the help of the toxins canavanine (*can1*Δ) and thialysine (*lyp1*Δ) for haploids but not specifically for *MAT***a** cells. Therefore, a mixture of cells of both mating types had to be used for the final mating test on a *MAT*α lawn. Regardless of the gene deletion, only the *MAT***a** cells out of this mixture were potentially able to mate with the tester strains. If, for unknown reasons, the fraction of *MAT*α cells was especially high or low, this would lead to a reduced or improved growth on a *MAT*α lawn compared to an average distribution of both mating types. Due to this, some strains from the deletion library might be selected for the top-ranking candidate list or sorted out regardless whether the respective factor affected *HML-SS* ΔI silencing or not. This imponderability only arises in the modified silencing screen. For the original synthetic lethality screen that searches for combinations of viable mutations that cause a lethal phenotype (Hartman *et al.*, 2001; Kaelin, 2005), a selection for a special mating type is not required, and no subsequent mating tests are performed.

Although the yeast deletion library provides a powerful tool to analyse the effect of single gene deletions on an observed genotypic feature, it cannot encompass the whole genome. The *S. cerevisiae* genome comprises ~ 6300 genes, with ~ 5000 of them being non-essential (Giaever *et al.*, 2002; Winzeler *et al.*, 1999)) and thus contained in the deletion library. Having considered the circumstances described above, the 264 factors that had been chosen as primary candidates for detailed analyses represent about 5 % of the deletion library strains. Undoubtedly, only a small portion of these individual mutants is potentially be involved in regulating *HML* silencing. Some factors could be excluded from further experiments because the gene deletion caused mating defects with wild-type *HML*-E. This led to a reduction from 264 primary to 219 secondary candidates, which were used in a re-array of the genetic screen. Furthermore, these candidates were used carrying an *HML-SS* ΔI plasmid in a second, independent experiment.

It has to be noted, that amino acid biosynthesis genes such as *TRP1* were excluded from further analysis since metabolism genes are unlikely to affect *HM* silencing. These deletion mutants might have appeared as candidates due to the fact that the deficiency prevented growth on medium lacking the respective supplement during the selection procedure of the genetic screen.

The experiment with the deletion strains carrying an *HML-SS* ΔI plasmid and the repetition of the genetic screen resulted in two lists of the best ten secondary candidates and three additional deletions were added due to their identity and a suspected involvement in *HML* silencing (see results). Tetrad analysis candidates from the genetic screen and subsequent silencing assays revealed another problematic issue of the deletion library screen. Although all controllable genotypic markers were identical in the tests of the individual factors, the mating experiments showed a very heterogeneous mating pattern (Fig. 20, Fig. 23), which impeded reliable conclusions about the involvement of these factors in *HML* silencing. This might be because the deletion strains possess several other genotypic differences next to the respective gene deletion. An explanation for this variance is the fact that the library was constructed over a long period in various laboratories. These differences could cause changes in the mating ability, cell growth or silencing, they segregate during the genetic crosses and cannot be identified later because they are not linked to auxotrophic markers or toxin resistances.

Another problematic issue of the yeast deletion library is indicated by the fact that the barcode analyses of some at first putative factors influencing *HML* silencing revealed that the identity of the deletions was not in all cases in agreement with the listed position in the microplates.

In addition to the genetic screen, the influence of some factors on *HML-SS* ΔI silencing was directly tested. Sir1 and Dot1 have been identified to influence silencing in a sensitized background using a candidate gene approach. In order to identify more novel factors that contribute to *HML* silencing, several experimental strategies are conceivable. Instead of using the deletion library, a *MATa HML-SS* ΔI strain could be used for transposon-mutagenesis and the resulting mutants could be analysed for reduced mating ability. The advantage of this method is that the positions of the random mutations are marked by the transposon insertion. The disadvantage is that transposons cause deletions or null alleles that will only affect non-essential genes or the promoters of essential genes in the genome. This means that only the effect of mutations of the ~ 5000 genes that are not required for viability and that are contained in the deletion library and in addition intergenic regions can be analysed by this method.

However, not only mutations that cause a silencing defect would be obtained, but also such that abrogate mating ability. To discriminate between an influence on *HML-SS* ΔI silencing and on mating ability in general, it would be possible to have *URA3* under the control of *HML*-E. Here, silencing of the reporter could be analysed independently of mating tests, for instance by comparing cell growth of serial dilutions of wild-type and mutant strains on 5-FOA and minimal medium lacking uracil.

Alternatively, with UV-mutagenesis, sequences in the whole genome could be randomly mutagenized. Here, the ~ 1000 essential genes might also be affected, and some might cause derepression of the synthetic *HML*-E allele. Disadvantageous is the fact that random UV-mutagenesis does not allow an easy identification. For any candidates that influence *HML* silencing, a more complicated identification of the mutated region would follow in order to identify the respective factor. To this end, the yeast genomic DNA library could be used for cloning by complementation.

To avoid gaining a high percentage of factors already known for their effect in silencing such as the Sir proteins in this new assay, it will be important to design the experiment in a way that only those factors that influence silencing of the minimal *HML*-E allele, but not *HM* silencing in general are obtained. Using an *HML-SS* ΔI plasmid instead of a genomic minimal silencer could circumvent this problem. If the observed α mating defect disappears after replacement of the *HML-SS* ΔI plasmid by an empty vector, the respective mutation affects *HML-SS* ΔI silencing specifically. In contrast, if plasmid replacement has no effect, *HML* and *HMR* silencing in general would be impaired by this mutation, by giving rise to an **a**/α phenotype.

If a limited number of factors is expected to be overrepresented among the candidates of a mutagenesis screen such as *sir2Δ*, *sir3Δ* or *sir4Δ*, a transformation with a collection of plasmids with wild-type copies of these genes might be considered. In case the candidates lose the defect in mating ability after the transformation, the effect was due to the already known factor. If not, probably a novel mutation had impaired *HML-SS* ΔI silencing.

In addition, it remains to be determined whether a *matΔ* strain provides a further sensitized background and thus would be more suited for these kinds of experiments. This is because silencing defects due to mutations in silencing elements do not necessarily provoke severe phenotypic changes. In contrary, mutations such as *sum1Δ* that reduce *HML-SS* ΔI silencing function might only cause slight alterations in non-quantitative analyses compared to

HML-SS ΔI alone. Indeed, this is even more problematic when not a distinct number of known mutations is tested, but a broad variety of unknown mutations as it would be the case with UV- or transposon-mutagenesis assays. Here, potentially small differences in mating ability might be masked by all other simultaneously tested mutants that have no effect.

Therefore, a sensitization of the experimental approach would be beneficial, which might be achieved by using a *mat*Δ strain. The *MATa1* promoter is deleted in these cells, which thus mate like a *MATa* strain. If α mating-type information becomes available in this strain due to *HML* derepression, a phenotypic switch from an **a** to an α phenotype takes place. Conversely, derepression of both *HMRa* in addition to *HML-SS* ΔI, as would be the case for *SIR* deletions, results in an **a**/α phenotype. Astonishingly, in an initial approach to analyse the applicability of a *mat*Δ strain, such a strain containing the minimal *HML-SS* ΔI silencer showed a stronger mating on a *MATa* lawn than it was previously shown for a strain with *HMR-SS*α (Kamakaka & Rine, 1998) (see also Fig. 14). This was surprising because the *HML-SS* silencer exhibited 60 % of the wild-type *HML*-E function, whereas the synthetic *HMR-SS* silencer only has 15 % of wild-type *HMR*-E (McNally & Rine, 1991).

Regardless which experimental approach will be followed in the future, the synthetic *HML*-E silencer generated during this study will likely allow the identification of additional factors that contribute to *HML* silencing and thus will allow novel insights into the mechanism of *HML* silencing.

4.4 Function of the Sum1/Rfm1/Hst1 deacetylation complex in replication initiation

The experiments performed to construct and analyse a synthetic, minimal *HML*-E silencer provided important new information to understand the role of Rap1, ORC and Sum1 function in *HML* silencing. The contribution of two of these factors, namely ORC and Sum1, as part of a complex with the histone deacetylase Hst1, in the regulation of origin activity was investigated in another major part of this thesis dealing with replication initiation.

Replication initiation in eukaryotic cells takes place on the chromatin template, which raises the question how the modification state of histones influences initiation. In this study, it was found that the DNA binding factor Sum1 recruited the histone deacetylase Hst1 within the Sum1/Rfm1/Hst1 complex to selected yeast origins to deacetylate histone H4. The most prominent effect was found for deacetylation of H4 K5, whereas other H4 acetylation sites were only affected at a minority of the origins.

How is histone deacetylation beneficial for initiation? The deacetylation of H4 K5 may help to stabilize the position of nucleosomes around the origin, for instance by altering DNA-histone contacts in the nucleosome, which has been shown to be important for initiation efficiency (Lipford & Bell, 2001). Alternatively, a particular deacetylated histone residue may recruit a histone-binding protein (complex) that recognizes H4 K5 in the deacetylated state and that has a positive effect on initiation. For instance, the acetylation may serve to recruit a chromatin remodeller that places the surrounding nucleosomes at a position that is conducive to initiation, which is in line with the observation of a role for chromatin remodellers in initiation (Flanagan & Peterson, 1999). It is also possible that Sum1/Rfm1/Hst1 indirectly affects initiation by altering the expression of genes encoding replication proteins.

It is interesting to note that histone deacetylation by Hst1 was found to be beneficial for initiation, whereas the HDAC Sir2 seems to negatively regulate initiation (Crampton *et al.*, 2008; Pappas *et al.*, 2004). Pappas *et al.* identified several origins whose plasmid maintenance capacity improved upon deletion of *SIR2*, while a decreased maintenance rate was found for a different set of origins in the absence of Hst1. Also, both *hst1Δ* and *sir2Δ* affected the survival of mutations in genes encoding replication factors, but *hst1Δ* reduced their viability (Fig. 24), whereas *sir2Δ* enhanced it (Crampton *et al.*, 2008; Pappas *et al.*, 2004). This indicates that the two HDACs both have a global effect on replication initiation, but that they influence initiation in opposite directions. This disparity may be explained by the different substrate specificities of the two enzymes. Here it is shown that Hst1 mainly deacetylates H4 K5, while the main target of Sir2 is H4 K16 (Imai *et al.*, 2000). The difference in specificity may lead to the recruitment of separate sets of regulatory factors that have different effects on nucleosome positioning and initiation. Thus, the effect of histone acetylation on efficiency of a particular origin seems to be highly dependent on the chromatin context of the origin. This is comparable to the effect of chromatin remodellers on transcription, where remodelling at one promoter can lead to the exposure of a transcription factor binding site, and thus to enhanced transcription, whereas remodelling at another promoter may lead to the occlusion of a binding site, and hence reduced expression of that gene (Becker & Horz, 2002).

It is also possible that the Sum1/Rfm1/Hst1 complex affects the time during S phase when an origin becomes active, much like increased histone acetylation at an origin by targeted Gcn5 or by the absence of Rpd3 advances initiation (Aparicio *et al.*, 2004; Vogelauer *et al.*, 2002). In this respect, *sum1Δ* and *hst1Δ* may delay the firing of many origins, such that the origins are inactivated by replication forks emanating from earlier origins and thus decreased in their firing efficiency, leading to synthetic lethality in orc2-1 cells.

A further possibility is that the Sum1 complex influences initiation via an effect on transcription of the neighbouring gene, since transcription has previously been shown to influence origin firing (Snyder et al., 1988).

Although Hst1 is an HDAC, it is also conceivable that deacetylation of a non-histone target, for instance a pre-RC component or another regulator of initiation, has an impact on initiation. However, the observation that mutation of the acetylatable lysine residues in H4 caused a similar effect on initiation as the deletion of *HST1* lends support to the notion that Hst1 affected initiation through histone deacetylation. As the same mutations that mimic an acetylated state, did not cause a synthetic lethality with a mutation in the ORC complex, *orc2-1*, this supports the suggestion that not all origins are influenced by histone H4 lysine acetylation. It can be concluded that although replication initiation is impaired at some origins due to the histone mutations, this effect is not severe enough to cause inviability if additionally ORC function is compromised. Therefore, effects of Hst1 beyond H4 deacetylation seem to account for the synthetic lethality of *orc2-1 hst1Δ* cells.

4.5 Did Sum1 affect sister chromatid cohesion?

The fact that the Sum1/Rfm1/Hst1 complex becomes essential when ORC function is compromised by the *orc2-1* mutation implies that a considerable number of origins in the yeast genome require this complex for initiation. One might then postulate that replication of the genome is not efficient enough in the absence of the Sum1 complex to support viability when initiation is reduced by an *orc* mutation. However, in this analysis only seven origins were identified as being regulated by this mechanism, a number that seems insufficient to explain the synthetic lethal effect. One possibility is that more Sum1-regulated origins exist that have not been identified in this analysis. Perhaps a re-analysis of the Sum1 localization data yields new Sum1 binding sites, as was the case for the re-evaluation of estrogen receptor binding data (Lupien et al., 2008).

Alternatively, the Sum1 complex may have additional functions that affect a second pathway parallel to ORC function. Next to its role in replication initiation, ORC also has a role in sister chromatid cohesion in that it mediates the interaction of sister chromatids in a pathway parallel to the interaction mediated by cohesin complexes (Shimada & Gasser, 2007; Suter *et al.*, 2004). Therefore, one explanation for the lethality between *sum1Δ/ rfm1Δ/ hst1Δ* and *orc2-1* is that it reflects an additional role for the Sum1/Rfm1/Hst1 complex in sister chromatid cohesion. However, since *sum1Δ* cells show no defect in sister chromatid cohesion (Shimada & Gasser, 2007), this Sum1-mediated cohesion pathway may act as a back-up in cells where ORC-mediated cohesion is impaired. In order to gain more information about this possible link, two different cohesion mutants were combined with *sum1Δ* to analyse whether the deletion of Sum1 has an effect on sister chromatid cohesion in a sensitized background. First the influence of *sum1Δ* on an *smc3-42* mutation was investigated. While a *SMC3* deletion is lethal because this gene is required for proliferation (Michaelis *et al.*, 1997), in a viable *smc3-42* strain, the kinetochores separate earlier than in wild-type cells, which results in a defect in holding sister chromatids together (Michaelis *et al.*, 1997). The experiments performed in this study showed that an *smc3-42 sum1Δ* mutation was not synthetically lethal but caused a slight synthetic growth defect. If *sum1Δ* contributed to sister chromatid cohesion it could be assumed that it caused a minor growth decrease of this temperature sensitive mutation because in the double mutants the thermosensitive phenotype was slightly enhanced. The next cohesion mutant analysed was *ctf18Δ*, since *orc2-1* has been shown to interact genetically with *ctf18Δ* (Suter *et al.*, 2004). Ctf18 is a component of an alternative replication factor C complex (Mayer *et al.*, 2001) and is required for sister chromatid cohesion (Hanna *et al.*, 2001). *CTF18* is not essential, but *ctf18Δ* causes chromosome instability and strong preanaphase delay (Kouprina *et al.*, 1994). Although *ctf18Δ sum1Δ* double mutants showed a delayed growth compared to either single mutant, again a deletion of *SUM1* did not cause a synthetic lethality with a cohesion mutant. These results suggested that Sum1, unlike ORC, does not play a major role in the regulation of sister chromatid cohesion, but that *sum1Δ* can impair cell growth when sister chromatid cohesion is already compromised.

4.6 Identification of additional factors with DNA-binding proximal to ORC

While ORC is the essential component of every yeast origin, Sum1 has been found to be co-localized with ORC only at some intergenic regions. Furthermore, the contribution of several auxiliary factors such as Sum1 or Rap1 to ORC's functions in replication initiation and silencing is required. However, it can be postulated that other DNA-binding factors exist that have an auxiliary role in replication initiation. Therefore, it was interesting to search for other factors that bind in the vicinity to ORC at multiple chromosomal locations because they constitute candidates for novel initiation auxiliary factors. Bioinformatics analyses for a co-localization of protein binding with ORC in intergenic regions revealed that ORC binds to

DNA sequences together with Sum1, and the analysis suggested also with several other factors. The most probable factors for this kind of co-occurrence were Gat3, Dat1 and Rgm1 (T. Manke, personal communication 2005). Interestingly, all three factors showed a synthetic lethality upon deletion in combination with a mutation in ORC, *orc2-1*. Since either single gene deletion was viable this suggested, that Dat1, Gat3 and Rgm1 play a role in a cellular process that also involves ORC. Since ORC exhibits a dual role in replication and silencing (Rehman & Yankulov, 2009) it seemed possible that Dat1, Gat3 and Rgm1 contributed to silencing or replication function. The fact that many intergenic regions with co-occurrence in ORC and Gat3 / Dat1 binding were located near the telomeres or the *HM* loci on the one hand, or on chromosome X on the other hand, supported this idea. However, several experiments performed during the course of this study indicated that neither telomeric / *HML* silencing nor initiation activity of chromosome X replication origins was significantly affected by *dat1*Δ or *gat3*Δ.

Since this was surprising, the ChIP-chip data, which had proposed a co-localization with ORC, were re-evaluated (T. Manke, personal communication 2007). A problematic issue was that Gat3 and Dat1 bind to AT-rich sequences, which are common at the telomeres. This might have led to the assumption that those factors have a special function at the telomeres, although the co-localization is coincidental and not due to the involvement in the same cellular mechanism. Furthermore, ChIP-chip analysis is more challenging for AT-rich regions, because the hybridisation efficiency is lower than at the rest of the genome (Euskirchen *et al.*, 2007). For these reasons, bioinformatics analysis was repeated (performed by T. Manke), but this time with exclusion of the telomeres. Surprisingly, the list of factors with high probability of binding proximal to ORC (Table 16) contained different factors than the original list, which had proposed Gat3, Dat1 and Rgm1 as high-ranking candidates. Furthermore, depending on the calculation method, only two factors (Yox1 and Hir2) are contained in both of the new lists, and these are not the top-ranked ones but on the positions seven to nine. Therefore, a wide variety of putative factors binding proximal to ORC, which depended on the underlying bioinformatics analysis of the available ChIP-chip data, existed. Considering that, it was difficult to infer, which of the gene products would be promising candidates for replication initiation factors. Therefore, no factors other than Dat1, Gat3 and Rgm1 were analysed. As all three tested factors (Dat1, Gat3 and Rgm1) showed a synthetic lethality together with *orc2-1*, while single deletion mutants are viable, a connection between ORC and Dat1 / Gat3 / Rgm1 in some kind of biological pathway seems very likely. This is partly supported by the calculated co-occurrences for binding at intergenic regions. Although a minor effect on telomeric silencing was observed in double mutants, no indication for an involvement of these factors in replication initiation was found, and the cause for the synthetic lethality remains unkown.

4.7 Summary of the main results of this study

During the course of this study, the role of the origin recognition complex (ORC) and of Sum1 in the processes of *HM* silencing and replication initiation was investigated. A synthetic, minimal *HML*-E silencer (*HML-SS* ΔI) that lacks the redundancy of wild-type *HML*-E was constructed. The requirement of Rap1, ORC and Sum1 binding sites within this construct for *HML-SS* ΔI silencing was demonstrated by analyzing mutations *in cis* and *in trans*. With Sir1 and Dot1, two factors were identified that contribute to *HML* silencing in this sensitized background.

Furthermore, the contribution of Sum1 as part of a HDAC with the histone deacetylase Hst1 to the initiation activity of a subset of origins was analysed. It was shown that histone H4 lysine 5 is the major target of Hst1, because *sum1*Δ and *hst1*Δ strains were significantly enriched in acetylation at this site compared to a wild-type strain. Interestingly, a mutation in the ORC complex (*orc2-1*) was found to be synthetically lethal with deletions of any of the three components of the Sum1/Rfm1/Hst1 HDAC as well as with deletions of three other factors (Dat1, Gat3, Rgm1) that are proposed to bind to intergenic regions in the vicinity of ORC.

References

Abdurashidova, G., Deganuto, M., Klima, R., Riva, S., Biamonti, G., Giacca, M. & Falaschi, A. (2000). Start sites of bidirectional DNA synthesis at the human lamin B2 origin. *Science* **287**, 2023-2026.

Abraham, J., Feldman, J., Nasmyth, K. A., Strathern, J. N., Klar, A. J., Broach, J. R. & Hicks, J. B. (1983). Sites required for position-effect regulation of mating-type information in yeast. *Cold Spring Harb Symp Quant Biol* **47 Pt 2**, 989-998.

Ahn, S. H., Cheung, W. L., Hsu, J. Y., Diaz, R. L., Smith, M. M. & Allis, C. D. (2005). Sterile 20 kinase phosphorylates histone H2B at serine 10 during hydrogen peroxide-induced apoptosis in S. cerevisiae. *Cell* **120**, 25-36.

Aladjem, M. I., Rodewald, L. W., Kolman, J. L. & Wahl, G. M. (1998). Genetic dissection of a mammalian replicator in the human beta-globin locus. *Science* **281**, 1005-1009.

Aparicio, J. G., Viggiani, C. J., Gibson, D. G. & Aparicio, O. M. (2004). The Rpd3-Sin3 histone deacetylase regulates replication timing and enables intra-S origin control in Saccharomyces cerevisiae. *Mol Cell Biol* **24**, 4769-4780.

Avery, O. T., Macleod, C. M. & McCarty, M. (1944). Studies on the Chemical Nature of the Substance Inducing Transformation of Pneumococcal Types : Induction of Transformation by a Desoxyribonucleic Acid Fraction Isolated from Pneumococcus Type Iii. *J Exp Med* **79**, 137-158.

Avner, P. & Heard, E. (2001). X-chromosome inactivation: counting, choice and initiation. *Nat Rev Genet* **2**, 59-67.

Baudin, A., Ozier-Kalogeropoulos, O., Denouel, A., Lacroute, F. & Cullin, C. (1993). A simple and efficient method for direct gene deletion in Saccharomyces cerevisiae. *Nucleic Acids Res* **21**, 3329-3330.

Becker, P. B. & Horz, W. (2002). ATP-dependent nucleosome remodeling. *Annu Rev Biochem* **71**, 247-273.

Bedalov, A., Hirao, M., Posakony, J., Nelson, M. & Simon, J. A. (2003). NAD+-dependent deacetylase Hst1p controls biosynthesis and cellular NAD+ levels in Saccharomyces cerevisiae. *Mol Cell Biol* **23**, 7044-7054.

Bell, S. P. & Dutta, A. (2002). DNA replication in eukaryotic cells. *Annu Rev Biochem* **71**, 333-374.

Berger, S. L. (2002). Histone modifications in transcriptional regulation. *Curr Opin Genet Dev* **12**, 142-148.

Blow, J. J. & Dutta, A. (2005). Preventing re-replication of chromosomal DNA. *Nat Rev Mol Cell Biol* **6**, 476-486.

Blow, J. J. & Ge, X. Q. (2008). Replication forks, chromatin loops and dormant replication origins. *Genome Biol* **9**, 244.

Boscheron, C., Maillet, L., Marcand, S., Tsai-Pflugfelder, M., Gasser, S. M. & Gilson, E. (1996). Cooperation at a distance between silencers and proto-silencers at the yeast HML locus. *Embo J* **15**, 2184-2195.

Brachmann, C. B., Sherman, J. M., Devine, S. E., Cameron, E. E., Pillus, L. & Boeke, J. D. (1995). The SIR2 gene family, conserved from bacteria to humans, functions in silencing, cell cycle progression, and chromosome stability. *Genes Dev* **9**, 2888-2902.

Brand, A. H., Breeden, L., Abraham, J., Sternglanz, R. & Nasmyth, K. (1985). Characterization of a "silencer" in yeast: a DNA sequence with properties opposite to those of a transcriptional enhancer. *Cell* **41**, 41-48.

Brewer, B. J. (1994). Intergenic DNA and the sequence requirements for replication initiation in eukaryotes. *Curr Opin Genet Dev* **4**, 196-202.

Chang, V. K., Donato, J. J., Chan, C. S. & Tye, B. K. (2004). Mcm1 promotes replication initiation by binding specific elements at replication origins. *Mol Cell Biol* **24**, 6514-6524.

Clyne, R. K. & Kelly, T. J. (1995). Genetic analysis of an ARS element from the fission yeast Schizosaccharomyces pombe. *Embo J* **14**, 6348-6357.

Conrad, M. N., Wright, J. H., Wolf, A. J. & Zakian, V. A. (1990). RAP1 protein interacts with yeast telomeres in vivo: overproduction alters telomere structure and decreases chromosome stability. *Cell* **63**, 739-750.

Contreras, A., Hale, T. K., Stenoien, D. L., Rosen, J. M., Mancini, M. A. & Herrera, R. E. (2003). The dynamic mobility of histone H1 is regulated by cyclin/CDK phosphorylation. *Mol Cell Biol* **23**, 8626-8636.

Costa, S. & Blow, J. J. (2007). The elusive determinants of replication origins. *EMBO Rep* **8**, 332-334.

Courbet, S., Gay, S., Arnoult, N., Wronka, G., Anglana, M., Brison, O. & Debatisse, M. (2008). Replication fork movement sets chromatin loop size and origin choice in mammalian cells. *Nature* **455**, 557-560.

Cox, K. H., Pinchak, A. B. & Cooper, T. G. (1999). Genome-wide transcriptional analysis in S. cerevisiae by mini-array membrane hybridization. *Yeast* **15**, 703-713.

Crampton, A., Chang, F., Pappas, D. L., Jr., Frisch, R. & Weinreich, M. (2008). An ARS Element Inhibits DNA Replication through a SIR2-Dependent Mechanism. *Mol Cell* **30**, 156-166.

Crick, F. H., Barnett, L., Brenner, S. & Watts-Tobin, R. J. (1961). General nature of the genetic code for proteins. *Nature* **192**, 1227-1232.

DeMarini, D. J., Creasy, C. L., Lu, Q., Mao, J., Sheardown, S. A., Sathe, G. M. & Livi, G. P. (2001). Oligonucleotide-mediated, PCR-independent cloning by homologous recombination. *Biotechniques* **30**, 520-523.

DePamphilis, M. L. (2005). Cell cycle dependent regulation of the origin recognition complex. *Cell Cycle* **4**, 70-79.

DePamphilis, M. L., Blow, J. J., Ghosh, S., Saha, T., Noguchi, K. & Vassilev, A. (2006). Regulating the licensing of DNA replication origins in metazoa. *Curr Opin Cell Biol* **18**, 231-239.

Derbyshire, M. K., Weinstock, K. G. & Strathern, J. N. (1996). HST1, a new member of the SIR2 family of genes. *Yeast* **12**, 631-640.

Diffley, J. F. & Stillman, B. (1988). Purification of a yeast protein that binds to origins of DNA replication and a transcriptional silencer. *Proc Natl Acad Sci U S A* **85**, 2120-2124.

Dijkwel, P. A., Wang, S. & Hamlin, J. L. (2002). Initiation sites are distributed at frequent intervals in the Chinese hamster dihydrofolate reductase origin of replication but are used with very different efficiencies. *Mol Cell Biol* **22**, 3053-3065.

Dinant, C., Houtsmuller, A. B. & Vermeulen, W. (2008). Chromatin structure and DNA damage repair. *Epigenetics Chromatin* **1**, 9.

Ehrenhofer-Murray, A. E., Rivier, D. H. & Rine, J. (1997). The role of Sas2, an acetyltransferase homologue of Saccharomyces cerevisiae, in silencing and ORC function. *Genetics* **145**, 923-934.

Ehrenhofer-Murray, A. E. (2004). Chromatin dynamics at DNA replication, transcription and repair. *Eur J Biochem* **271**, 2335-2349.

Ekwall, K. (2005). Genome-wide analysis of HDAC function. *Trends Genet* **21**, 608-615.

ENCODE (2004). The ENCODE (ENCyclopedia Of DNA Elements) Project. *Science* **306**, 636-640.

Estruch, F. (1991). The yeast putative transcriptional repressor RGM1 is a proline-rich zinc finger protein. *Nucleic Acids Res* **19**, 4873-4877.

Euskirchen, G. M., Rozowsky, J. S., Wei, C. L. & other authors (2007). Mapping of transcription factor binding regions in mammalian cells by ChIP: comparison of array- and sequencing-based technologies. *Genome Res* **17**, 898-909.

Felsenfeld, G. & Groudine, M. (2003). Controlling the double helix. *Nature* **421**, 448-453.

Ferguson, B. M., Brewer, B. J., Reynolds, A. E. & Fangman, W. L. (1991). A yeast origin of replication is activated late in S phase. *Cell* **65**, 507-515.

Flanagan, J. F. & Peterson, C. L. (1999). A role for the yeast SWI/SNF complex in DNA replication. *Nucleic Acids Res* **27**, 2022-2028.

Foss, M., McNally, F. J., Laurenson, P. & Rine, J. (1993). Origin recognition complex (ORC) in transcriptional silencing and DNA replication in S. cerevisiae. *Science* **262**, 1838-1844.

Fox, C. A., Loo, S., Dillin, A. & Rine, J. (1995). The origin recognition complex has essential functions in transcriptional silencing and chromosomal replication. *Genes Dev* **9**, 911-924.

Fox, C. A., Ehrenhofer-Murray, A. E., Loo, S. & Rine, J. (1997). The origin recognition complex, SIR1, and the S phase requirement for silencing. *Science* **276**, 1547-1551.

Freiman, R. N. & Tjian, R. (2003). Regulating the regulators: lysine modifications make their mark. *Cell* **112**, 11-17.

Gardner, K. A., Rine, J. & Fox, C. A. (1999). A region of the Sir1 protein dedicated to recognition of a silencer and required for interaction with the Orc1 protein in saccharomyces cerevisiae. *Genetics* **151**, 31-44.

Giaever, G., Chu, A. M., Ni, L. & other authors (2002). Functional profiling of the Saccharomyces cerevisiae genome. *Nature* **418**, 387-391.

Gilbert, D. M. (2001). Making sense of eukaryotic DNA replication origins. *Science* **294**, 96-100.

Gilson, E., Roberge, M., Giraldo, R., Rhodes, D. & Gasser, S. M. (1993). Distortion of the DNA double helix by RAP1 at silencers and multiple telomeric binding sites. *J Mol Biol* **231**, 293-310.

Gottschling, D. E., Aparicio, O. M., Billington, B. L. & Zakian, V. A. (1990). Position effect at S. cerevisiae telomeres: reversible repression of Pol II transcription. *Cell* **63**, 751-762.

Grant, P. A. & Berger, S. L. (1999). Histone acetyltransferase complexes. *Semin Cell Dev Biol* **10**, 169-177.

Hallstrom, B. M. & Janke, A. (2009). Gnathostome phylogenomics utilizing lungfish EST sequences. *Mol Biol Evol* **26**, 463-471.

Hanna, J. S., Kroll, E. S., Lundblad, V. & Spencer, F. A. (2001). Saccharomyces cerevisiae CTF18 and CTF4 are required for sister chromatid cohesion. *Mol Cell Biol* **21**, 3144-3158.

Harbison, C. T., Gordon, D. B., Lee, T. I. & other authors (2004). Transcriptional regulatory code of a eukaryotic genome. *Nature* **431**, 99-104.

Hartman, J. L. t., Garvik, B. & Hartwell, L. (2001). Principles for the buffering of genetic variation. *Science* **291**, 1001-1004.

Hecht, A., Laroche, T., Strahl-Bolsinger, S., Gasser, S. M. & Grunstein, M. (1995). Histone H3 and H4 N-termini interact with SIR3 and SIR4 proteins: a molecular model for the formation of heterochromatin in yeast. *Cell* **80**, 583-592.

Herskowitz, I., Rine, J. & Strathern, J. N. (1992). Mating-type determination and matingtype interconversion in Saccharomyces cerevisiae. In *The Molecular and Cellular Biology of the Yeast Saccharomyces*. Edited by E. W. Jones, J. R. Pringle & J. R. Broach. Cold Spring Harbor, NY: Cold Spring Harbor Laboratory Press.

Herskowitz, I. S., Hicks, J. B. & Rine, J. (1977). Mating type interconversion in yeast and its relationship to development in higher eucaryotes. In *ICN-UCLA Symposia on Molecular and Cellular Biology: Molecular Approaches to Eucaryotic Genetic Systems*, pp. 193-202. New York: Academic Press, New York.

Hoppe, G. J., Tanny, J. C., Rudner, A. D., Gerber, S. A., Danaie, S., Gygi, S. P. & Moazed, D. (2002). Steps in assembly of silent chromatin in yeast: Sir3-independent binding of a Sir2/Sir4 complex to silencers and role for Sir2-dependent deacetylation. *Mol Cell Biol* **22**, 4167-4180.

Iizuka, M. & Smith, M. M. (2003). Functional consequences of histone modifications. *Curr Opin Genet Dev* **13**, 154-160.

Imai, S., Armstrong, C. M., Kaeberlein, M. & Guarente, L. (2000). Transcriptional silencing and longevity protein Sir2 is an NAD-dependent histone deacetylase. *Nature* **403**, 795-800.

Ina, S., Sasaki, T., Yokota, Y. & Shinomiya, T. (2001). A broad replication origin of Drosophila melanogaster, oriDalpha, consists of AT-rich multiple discrete initiation sites. *Chromosoma* **109**, 551-564.

Irlbacher, H., Franke, J., Manke, T., Vingron, M. & Ehrenhofer-Murray, A. E. (2005). Control of replication initiation and heterochromatin formation in Saccharomyces cerevisiae by a regulator of meiotic gene expression. *Genes Dev* **19**, 1811-1822.

Jenuwein, T. & Allis, C. D. (2001). Translating the histone code. *Science* **293**, 1074-1080.

Jiang, C. & Pugh, B. F. (2009). Nucleosome positioning and gene regulation: advances through genomics. *Nat Rev Genet* **10**, 161-172.

Kaelin, W. G., Jr. (2005). The concept of synthetic lethality in the context of anticancer therapy. *Nat Rev Cancer* **5**, 689-698.

Kamakaka, R. T. & Rine, J. (1998). Sir- and silencer-independent disruption of silencing in Saccharomyces by Sas10p. *Genetics* **149**, 903-914.

Kim, J. H. & Workman, J. L. (2010). Histone acetylation in heterochromatin assembly. *Genes Dev* **24**, 738-740.

Kimmerly, W., Buchman, A., Kornberg, R. & Rine, J. (1988). Roles of two DNA-binding factors in replication, segregation and transcriptional repression mediated by a yeast silencer. *Embo J* **7**, 2241-2253.

Kimura, A., Umehara, T. & Horikoshi, M. (2002). Chromosomal gradient of histone acetylation established by Sas2p and Sir2p functions as a shield against gene silencing. *Nat Genet* **32**, 370-377.

Kingston, R. E. & Narlikar, G. J. (1999). ATP-dependent remodeling and acetylation as regulators of chromatin fluidity. *Genes Dev* **13**, 2339-2352.

Klebe, R. J., Harriss, J. V., Sharp, Z. D. & Douglas, M. G. (1983). A general method for polyethylene-glycol-induced genetic transformation of bacteria and yeast. *Gene* **25**, 333-341.

Klein, F., Mahr, P., Galova, M., Buonomo, S. B., Michaelis, C., Nairz, K. & Nasmyth, K. (1999). A central role for cohesins in sister chromatid cohesion, formation of axial elements, and recombination during yeast meiosis. *Cell* **98**, 91-103.

Kornberg, R. D. & Lorch, Y. (1999). Chromatin-modifying and -remodeling complexes. *Curr Opin Genet Dev* **9**, 148-151.

Kouprina, N., Kroll, E., Kirillov, A., Bannikov, V., Zakharyev, V. & Larionov, V. (1994). CHL12, a gene essential for the fidelity of chromosome transmission in the yeast Saccharomyces cerevisiae. *Genetics* **138**, 1067-1079.

Kouzarides, T. (2002). Histone methylation in transcriptional control. *Curr Opin Genet Dev* **12**, 198-209.

Kouzarides, T. (2007). Chromatin modifications and their function. *Cell* **128**, 693-705.

Kurdistani, S. K., Robyr, D., Tavazoie, S. & Grunstein, M. (2002). Genome-wide binding map of the histone deacetylase Rpd3 in yeast. *Nat Genet* **31**, 248-254.

Lander, E. S., Linton, L. M., Birren, B. & other authors (2001). Initial sequencing and analysis of the human genome. *Nature* **409**, 860-921.

Lee, D. G. & Bell, S. P. (1997). Architecture of the yeast origin recognition complex bound to origins of DNA replication. *Mol Cell Biol* **17**, 7159-7168.

Lee, T. I., Rinaldi, N. J., Robert, F. & other authors (2002). Transcriptional regulatory networks in Saccharomyces cerevisiae. *Science* **298**, 799-804.

Li, R., Yu, D. S., Tanaka, M., Zheng, L., Berger, S. L. & Stillman, B. (1998). Activation of chromosomal DNA replication in Saccharomyces cerevisiae by acidic transcriptional activation domains. *Mol Cell Biol* **18**, 1296-1302.

Lipford, J. R. & Bell, S. P. (2001). Nucleosomes positioned by ORC facilitate the initiation of DNA replication. *Mol Cell* **7**, 21-30.

Little, R. D., Platt, T. H. & Schildkraut, C. L. (1993). Initiation and termination of DNA replication in human rRNA genes. *Mol Cell Biol* **13**, 6600-6613.

Loo, S., Fox, C. A., Rine, J., Kobayashi, R., Stillman, B. & Bell, S. (1995a). The origin recognition complex in silencing, cell cycle progression, and DNA replication. *Mol Biol Cell* **6**, 741-756.

Loo, S., Laurenson, P., Foss, M., Dillin, A. & Rine, J. (1995b). Roles of ABF1, NPL3, and YCL54 in silencing in Saccharomyces cerevisiae. *Genetics* **141**, 889-902.

Loo, S. & Rine, J. (1995). Silencing and heritable domains of gene expression. *Annu Rev Cell Dev Biol* **11**, 519-548.

Luger, K., Rechsteiner, T. J. & Richmond, T. J. (1999). Expression and purification of recombinant histones and nucleosome reconstitution. *Methods Mol Biol* **119**, 1-16.

Luo, K., Vega-Palas, M. A. & Grunstein, M. (2002). Rap1-Sir4 binding independent of other Sir, yKu, or histone interactions initiates the assembly of telomeric heterochromatin in yeast. *Genes Dev* **16**, 1528-1539.

Lupien, M., Eeckhoute, J., Meyer, C. A., Wang, Q., Zhang, Y., Li, W., Carroll, J. S., Liu, X. S. & Brown, M. (2008). FoxA1 translates epigenetic signatures into enhancer-driven lineage-specific transcription. *Cell* **132**, 958-970.

Lustig, A. J., Kurtz, S. & Shore, D. (1990). Involvement of the silencer and UAS binding protein RAP1 in regulation of telomere length. *Science* **250**, 549-553.

Lynch, P. J., Fraser, H. B., Sevastopoulos, E., Rine, J. & Rusche, L. N. (2005). Sum1p, the origin recognition complex, and the spreading of a promoter-specific repressor in Saccharomyces cerevisiae. *Mol Cell Biol* **25**, 5920-5932.

Mahoney, D. J. & Broach, J. R. (1989). The HML mating-type cassette of Saccharomyces cerevisiae is regulated by two separate but functionally equivalent silencers. *Mol Cell Biol* **9**, 4621-4630.

Mahoney, D. J., Marquardt, R., Shei, G. J., Rose, A. B. & Broach, J. R. (1991). Mutations in the HML E silencer of Saccharomyces cerevisiae yield metastable inheritance of transcriptional repression. *Genes Dev* **5**, 605-615.

Marahrens, Y. & Stillman, B. (1992). A yeast chromosomal origin of DNA replication defined by multiple functional elements. *Science* **255**, 817-823.

Martino, F., Kueng, S., Robinson, P. & other authors (2009). Reconstitution of yeast silent chromatin: multiple contact sites and O-AADPR binding load SIR complexes onto nucleosomes in vitro. *Mol Cell* **33**, 323-334.

Mayer, M. L., Gygi, S. P., Aebersold, R. & Hieter, P. (2001). Identification of RFC(Ctf18p, Ctf8p, Dcc1p): an alternative RFC complex required for sister chromatid cohesion in S. cerevisiae. *Mol Cell* **7**, 959-970.

McCord, R., Pierce, M., Xie, J., Wonkatal, S., Mickel, C. & Vershon, A. K. (2003). Rfm1, a novel tethering factor required to recruit the Hst1 histone deacetylase for repression of middle sporulation genes. *Mol Cell Biol* **23**, 2009-2016.

McNally, F. J. & Rine, J. (1991). A synthetic silencer mediates SIR-dependent functions in Saccharomyces cerevisiae. *Mol Cell Biol* **11**, 5648-5659.

Megee, P. C., Morgan, B. A., Mittman, B. A. & Smith, M. M. (1990). Genetic analysis of histone H4: essential role of lysines subject to reversible acetylation. *Science* **247**, 841-845.

Meijsing, S. H. & Ehrenhofer-Murray, A. E. (2001). The silencing complex SAS-I links histone acetylation to the assembly of repressed chromatin by CAF-I and Asf1 in Saccharomyces cerevisiae. *Genes Dev* **15**, 3169-3182.

Michaelis, C., Ciosk, R. & Nasmyth, K. (1997). Cohesins: chromosomal proteins that prevent premature separation of sister chromatids. *Cell* **91**, 35-45.

Micklem, G., Rowley, A., Harwood, J., Nasmyth, K. & Diffley, J. F. (1993). Yeast origin recognition complex is involved in DNA replication and transcriptional silencing. *Nature* **366**, 87-89.

Muller, H. J. (1930). Types of visible variations induced by X-rays inDrosophila. *Journal of Genetics* **22**, 299-334.

Ng, H. H., Feng, Q., Wang, H., Erdjument-Bromage, H., Tempst, P., Zhang, Y. & Struhl, K. (2002). Lysine methylation within the globular domain of histone H3 by Dot1 is important for telomeric silencing and Sir protein association. *Genes Dev* **16**, 1518-1527.

Nguyen, V. Q., Co, C. & Li, J. J. (2001). Cyclin-dependent kinases prevent DNA re-replication through multiple mechanisms. *Nature* **411**, 1068-1073.

Nieduszynski, C. A., Knox, Y. & Donaldson, A. D. (2006). Genome-wide identification of replication origins in yeast by comparative genomics. *Genes Dev* **20**, 1874-1879.

Oki, M. & Kamakaka, R. T. (2002). Blockers and barriers to transcription: competing activities? *Curr Opin Cell Biol* **14**, 299-304.

Okuno, Y., Satoh, H., Sekiguchi, M. & Masukata, H. (1999). Clustered adenine/thymine stretches are essential for function of a fission yeast replication origin. *Mol Cell Biol* **19**, 6699-6709.

Osada, S., Sutton, A., Muster, N., Brown, C. E., Yates, J. R., 3rd, Sternglanz, R. & Workman, J. L. (2001). The yeast SAS (something about silencing) protein complex contains a MYST-type putative acetyltransferase and functions with chromatin assembly factor ASF1. *Genes Dev* **15**, 3155-3168.

Ozaydin, B. & Rine, J. (2009). Expanded roles of the origin recognition complex in the architecture and function of silenced chromatin in Saccharomyces cerevisiae. *Mol Cell Biol* **30**, 626-639.

Palacios DeBeer, M. A. & Fox, C. A. (1999). A role for a replicator dominance mechanism in silencing. *Embo J* **18**, 3808-3819.

Pappas, D. L., Jr., Frisch, R. & Weinreich, M. (2004). The NAD(+)-dependent Sir2p histone deacetylase is a negative regulator of chromosomal DNA replication. *Genes Dev* **18**, 769-781.

Paux, E., Sourdille, P., Salse, J. & other authors (2008). A physical map of the 1-gigabase bread wheat chromosome 3B. *Science* **322**, 101-104.

Petes, T. D. & Botstein, D. (1977). Simple Mendelian inheritance of the reiterated ribosomal DNA of yeast. *Proc Natl Acad Sci U S A* **74**, 5091-5095.

Pierce, M., Benjamin, K. R., Montano, S. P., Georgiadis, M. M., Winter, E. & Vershon, A. K. (2003). Sum1 and Ndt80 proteins compete for binding to middle sporulation element sequences that control meiotic gene expression. *Mol Cell Biol* **23**, 4814-4825.

Pijnappel, W. W., Schaft, D., Roguev, A. & other authors (2001). The S. cerevisiae SET3 complex includes two histone deacetylases, Hos2 and Hst1, and is a meiotic-specific repressor of the sporulation gene program. *Genes Dev* **15**, 2991-3004.

Pillus, L. & Rine, J. (1989). Epigenetic inheritance of transcriptional states in S. cerevisiae. *Cell* **59**, 637-647.

Planta, R. J., Goncalves, P. M. & Mager, W. H. (1995). Global regulators of ribosome biosynthesis in yeast. *Biochem Cell Biol* **73**, 825-834.

Pryde, F. E. & Louis, E. J. (1999). Limitations of silencing at native yeast telomeres. *Embo J* **18**, 2538-2550.

Raghuraman, M. K., Winzeler, E. A., Collingwood, D. & other authors (2001). Replication dynamics of the yeast genome. *Science* **294**, 115-121.

Redon, C., Pilch, D., Rogakou, E., Sedelnikova, O., Newrock, K. & Bonner, W. (2002). Histone H2A variants H2AX and H2AZ. *Curr Opin Genet Dev* **12**, 162-169.

Rehman, M. A. & Yankulov, K. (2009). The dual role of autonomously replicating sequences as origins of replication and as silencers. *Curr Genet* **55**, 357-363.

Reid, J. L., Iyer, V. R., Brown, P. O. & Struhl, K. (2000). Coordinate regulation of yeast ribosomal protein genes is associated with targeted recruitment of Esa1 histone acetylase. *Mol Cell* **6**, 1297-1307.

Reifsnyder, C., Lowell, J., Clarke, A. & Pillus, L. (1996). Yeast SAS silencing genes and human genes associated with AML and HIV-1 Tat interactions are homologous with acetyltransferases. *Nat Genet* **14**, 42-49.

Richmond, T. J. & Davey, C. A. (2003). The structure of DNA in the nucleosome core. *Nature* **423**, 145-150.

Rine, J. & Herskowitz, I. (1987). Four genes responsible for a position effect on expression from HML and HMR in Saccharomyces cerevisiae. *Genetics* **116**, 9-22.

Robert, F., Pokholok, D. K., Hannett, N. M., Rinaldi, N. J., Chandy, M., Rolfe, A., Workman, J. L., Gifford, D. K. & Young, R. A. (2004). Global position and recruitment of HATs and HDACs in the yeast genome. *Mol Cell* **16**, 199-209.

Roth, S. Y. (1995). Chromatin-mediated transcriptional repression in yeast. *Curr Opin Genet Dev* **5**, 168-173.

Rusche, L. N. & Rine, J. (2001). Conversion of a gene-specific repressor to a regional silencer. *Genes Dev* **15**, 955-967.

Rusche, L. N., Kirchmaier, A. L. & Rine, J. (2002). Ordered nucleation and spreading of silenced chromatin in Saccharomyces cerevisiae. *Mol Biol Cell* **13**, 2207-2222.

Rusche, L. N., Kirchmaier, A. L. & Rine, J. (2003). The establishment, inheritance, and function of silenced chromatin in Saccharomyces cerevisiae. *Annu Rev Biochem* **72**, 481-516.

Saha, A., Wittmeyer, J. & Cairns, B. R. (2006). Chromatin remodelling: the industrial revolution of DNA around histones. *Nat Rev Mol Cell Biol* **7**, 437-447.

Sambrook, J., Fritsch, E. F. & Maniatis, T. (1989). *Molecular Cloning: A Laboratory Manual*. Cold Spring Harbor: Cold Spring Harbor Laboratory press.

Sharp, J. A., Fouts, E. T., Krawitz, D. C. & Kaufman, P. D. (2001). Yeast histone deposition protein Asf1p requires Hir proteins and PCNA for heterochromatic silencing. *Curr Biol* **11**, 463-473.

Sherman, F. (1991). Getting started with yeast. *Methods Enzymol* **194**, 3-21.

Shimada, K. & Gasser, S. M. (2007). The origin recognition complex functions in sister-chromatid cohesion in Saccharomyces cerevisiae. *Cell* **128**, 85-99.

Shor, E., Warren, C. L., Tietjen, J. & other authors (2009). The origin recognition complex interacts with a subset of metabolic genes tightly linked to origins of replication. *PLoS Genet* **5**, e1000755.

Sif, S. (2004). ATP-dependent nucleosome remodeling complexes: enzymes tailored to deal with chromatin. *J Cell Biochem* **91**, 1087-1098.

Sikorski, R. S. & Hieter, P. (1989). A system of shuttle vectors and yeast host strains designed for efficient manipulation of DNA in Saccharomyces cerevisiae. *Genetics* **122**, 19-27.

Smith, C. M., Gafken, P. R., Zhang, Z., Gottschling, D. E., Smith, J. B. & Smith, D. L. (2003). Mass spectrometric quantification of acetylation at specific lysines within the amino-terminal tail of histone H4. *Anal Biochem* **316**, 23-33.

Smith, J. S. & Boeke, J. D. (1997). An unusual form of transcriptional silencing in yeast ribosomal DNA. *Genes Dev* **11**, 241-254.

Smith, M. M. (2002). Centromeres and variant histones: what, where, when and why? *Curr Opin Cell Biol* **14**, 279-285.

Snyder, M., Sapolsky, R. J. & Davis, R. W. (1988). Transcription interferes with elements important for chromosome maintenance in Saccharomyces cerevisiae. *Mol Cell Biol* **8**, 2184-2194.

Stavenhagen, J. B. & Zakian, V. A. (1994). Internal tracts of telomeric DNA act as silencers in Saccharomyces cerevisiae. *Genes Dev* **8**, 1411-1422.

Stillman, B. (2005). Origin recognition and the chromosome cycle. *FEBS Lett* **579**, 877-884.

Strahl, B. D. & Allis, C. D. (2000). The language of covalent histone modifications. *Nature* **403**, 41-45.

Strahl-Bolsinger, S., Hecht, A., Luo, K. & Grunstein, M. (1997). SIR2 and SIR4 interactions differ in core and extended telomeric heterochromatin in yeast. *Genes Dev* **11**, 83-93.

Straight, A. F., Shou, W., Dowd, G. J., Turck, C. W., Deshaies, R. J., Johnson, A. D. & Moazed, D. (1999). Net1, a Sir2-associated nucleolar protein required for rDNA silencing and nucleolar integrity. *Cell* **97**, 245-256.

Suka, N., Luo, K. & Grunstein, M. (2002). Sir2p and Sas2p opposingly regulate acetylation of yeast histone H4 lysine16 and spreading of heterochromatin. *Nat Genet* **32**, 378-383.

Sussel, L. & Shore, D. (1991). Separation of transcriptional activation and silencing functions of the RAP1-encoded repressor/activator protein 1: isolation of viable mutants affecting both silencing and telomere length. *Proc Natl Acad Sci U S A* **88**, 7749-7753.

Sussel, L., Vannier, D. & Shore, D. (1993). Epigenetic switching of transcriptional states: cis- and trans-acting factors affecting establishment of silencing at the HMR locus in Saccharomyces cerevisiae. *Mol Cell Biol* **13**, 3919-3928.

Suter, B., Tong, A., Chang, M., Yu, L., Brown, G. W., Boone, C. & Rine, J. (2004). The origin recognition complex links replication, sister chromatid cohesion and transcriptional silencing in Saccharomyces cerevisiae. *Genetics* **167**, 579-591.

Sutton, A., Heller, R. C., Landry, J., Choy, J. S., Sirko, A. & Sternglanz, R. (2001). A novel form of transcriptional silencing by Sum1-1 requires Hst1 and the origin recognition complex. *Mol Cell Biol* **21**, 3514-3522.

The Chimpanzee Sequencing and Analysis Consortium (2005). Initial sequence of the chimpanzee genome and comparison with the human genome. *Nature* **437**, 69-87.

Toledo, F., Baron, B., Fernandez, M. A., Lachages, A. M., Mayau, V., Buttin, G. & Debatisse, M. (1998). oriGNAI3: a narrow zone of preferential replication initiation in mammalian cells identified by 2D gel and competitive PCR replicon mapping techniques. *Nucleic Acids Res* **26**, 2313-2321.

Tong, A. H., Evangelista, M., Parsons, A. B. & other authors (2001). Systematic genetic analysis with ordered arrays of yeast deletion mutants. *Science* **294**, 2364-2368.

Tong, A. H., Lesage, G., Bader, G. D. & other authors (2004). Global mapping of the yeast genetic interaction network. *Science* **303**, 808-813.

Tong, A. H. & Boone, C. (2007). High-Throughput Strain Construction and Systematic Synthetic Lethal Screening in Saccharomyces cerevisiae. In *Methods In Microbiology*, pp. 369-386, 706-707: Elsevier Science.

Tsankova, N., Renthal, W., Kumar, A. & Nestler, E. J. (2007). Epigenetic regulation in psychiatric disorders. *Nat Rev Neurosci* **8**, 355-367.

Urnov, F. D. & Wolffe, A. P. (2001). Chromatin remodeling and transcriptional activation: the cast (in order of appearance). *Oncogene* **20**, 2991-3006.

van Leeuwen, F., Gafken, P. R. & Gottschling, D. E. (2002). Dot1p modulates silencing in yeast by methylation of the nucleosome core. *Cell* **109**, 745-756.

van Welsem, T., Frederiks, F., Verzijlbergen, K. F., Faber, A. W., Nelson, Z. W., Egan, D. A., Gottschling, D. E. & van Leeuwen, F. (2008). Synthetic lethal screens identify gene silencing processes in yeast and implicate the acetylated amino terminus of Sir3 in recognition of the nucleosome core. *Mol Cell Biol* **28**, 3861-3872.

Vas, A., Mok, W. & Leatherwood, J. (2001). Control of DNA rereplication via Cdc2 phosphorylation sites in the origin recognition complex. *Mol Cell Biol* **21**, 5767-5777.

Vignali, M., Steger, D. J., Neely, K. E. & Workman, J. L. (2000). Distribution of acetylated histones resulting from Gal4-VP16 recruitment of SAGA and NuA4 complexes. *Embo J* **19**, 2629-2640.

Vogelauer, M., Rubbi, L., Lucas, I., Brewer, B. J. & Grunstein, M. (2002). Histone acetylation regulates the time of replication origin firing. *Mol Cell* **10**, 1223-1233.

Vujcic, M., Miller, C. A. & Kowalski, D. (1999). Activation of silent replication origins at autonomously replicating sequence elements near the HML locus in budding yeast. *Mol Cell Biol* **19**, 6098-6109.

Wach, A., Brachat, A., Pohlmann, R. & Philippsen, P. (1994). New heterologous modules for classical or PCR-based gene disruptions in Saccharomyces cerevisiae. *Yeast* **10**, 1793-1808.

Weber, J. M., Irlbacher, H. & Ehrenhofer-Murray, A. E. (2008). Control of replication initiation by the Sum1/Rfm1/Hst1 histone deacetylase. *BMC Mol Biol* **9**, 100.

Weinreich, M., Palacios DeBeer, M. A. & Fox, C. A. (2004). The activities of eukaryotic replication origins in chromatin. *Biochim Biophys Acta* **1677**, 142-157.

Wilmes, G. M. & Bell, S. P. (2002). The B2 element of the Saccharomyces cerevisiae ARS1 origin of replication requires specific sequences to facilitate pre-RC formation. *Proc Natl Acad Sci U S A* **99**, 101-106.

Wilmes, G. M., Archambault, V., Austin, R. J., Jacobson, M. D., Bell, S. P. & Cross, F. R. (2004). Interaction of the S-phase cyclin Clb5 with an "RXL" docking sequence in the initiator protein Orc6 provides an origin-localized replication control switch. *Genes Dev* **18**, 981-991.

Winston, F. & Carlson, M. (1992). Yeast SNF/SWI transcriptional activators and the SPT/SIN chromatin connection. *Trends Genet* **8**, 387-391.

Winter, E. & Varshavsky, A. (1989). A DNA binding protein that recognizes oligo(dA).oligo(dT) tracts. *Embo J* **8**, 1867-1877.

Winzeler, E. A., Shoemaker, D. D., Astromoff, A. & other authors (1999). Functional characterization of the S. cerevisiae genome by gene deletion and parallel analysis. *Science* **285**, 901-906.

Wright, J. H., Gottschling, D. E. & Zakian, V. A. (1992). Saccharomyces telomeres assume a non-nucleosomal chromatin structure. *Genes Dev* **6**, 197-210.

Wu, C., Bassett, A. & Travers, A. (2007). A variable topology for the 30-nm chromatin fibre. *EMBO Rep* **8**, 1129-1134.

Wu, J. & Grunstein, M. (2000). 25 years after the nucleosome model: chromatin modifications. *Trends Biochem Sci* **25**, 619-623.

Wyrick, J. J., Aparicio, J. G., Chen, T., Barnett, J. D., Jennings, E. G., Young, R. A., Bell, S. P. & Aparicio, O. M. (2001). Genome-wide distribution of ORC and MCM proteins in S. cerevisiae: high-resolution mapping of replication origins. *Science* **294**, 2357-2360.

Xie, J., Pierce, M., Gailus-Durner, V., Wagner, M., Winter, E. & Vershon, A. K. (1999). Sum1 and Hst1 repress middle sporulation-specific gene expression during mitosis in Saccharomyces cerevisiae. *Embo J* **18**, 6448-6454.

Zou, L. & Stillman, B. (2000). Assembly of a complex containing Cdc45p, replication protein A, and Mcm2p at replication origins controlled by S-phase cyclin-dependent kinases and Cdc7p-Dbf4p kinase. *Mol Cell Biol* **20**, 3086-3096.

Acknowledgements

Eingangs möchte ich Ann besonders herzlich für die aktuelle und interessante Themenstellung, ihre stetige Anteilnahme am Fortgang der Doktorarbeit sowie die wissenschaftlichen Anregungen und produktiven Diskussionen danken.

Ein großer Dank gilt den zuvor an Teilthemen beteiligten Personen ohne deren Arbeiten das Projekt in der vorliegenden Form nicht entstanden wäre.

Den Mitarbeiterinnen und Mitarbeitern der Arbeitsgruppe für Genetik gilt mein Dank für das ausgesprochen gute Arbeitsklima. Hervorhebend möchte ich Stefan für die vielen hilfreichen Tipps und Anke für ihre intensive Diskussionsbereitschaft danken. Bei Martin bedanke ich mich für die immer aktuellen Erörterungen des Weltgeschehens und die harmonische Teilung der Arbeitsfläche sowie bei Jessica für manchen Hefestamm und Inkubatorenwechsel. Christiane und Karolin sei für die hervorragende Organisation in unserem Labor gedankt, Martina für ihre unermüdliche Unterstützung und Rolf für seine herzliche Art, die ein oder andere Anekdote und die Freude an Chilis.
Maria und Tanja danke ich für die Ratschläge bezüglich des Manuskripts und die vielen gemeinschaftlichen Mittagspausen mit Gesine und Rita, die zu jeder Zeit ein offenes Ohr für alle Probleme hatte.

Meiner Familie und meinen engsten Freunden sei für den fortwährenden Zuspruch und das Leben außerhalb der Forschung, in einem ruhigen Moment persönlich gedankt.

Ein tiefer Dank gilt Toni, die immer an mich geglaubt hat.

Zuletzt danke ich Kerstin für die moralische Unterstützung, aufmunternden Worte und ihre Zeit innerhalb der letzten Jahre.

Publications

Heise F, Chung H, Weber JM, Xu Z, Klein-Hitpass L, Steinmetz LM, Vingron M, Ehrenhofer-Murray AE (2011) Genome-wide H4 K16 acetylation by SAS-I is deposited independently of transcription and histone exchange. *Nucleic Acids Res* **40** (1):65-74

Ehrentraut S, Hassler M, Oppikofer M, Kueng S, Weber JM, Gasser SM, Ladurner AG, Ehrenhofer-Murray AE (2011). Structural basis for the role of the Sir3 AAA+ domain in silencing: interaction with Sir4 and unmethylated histone H3K79. *Genes Dev.* **25(17)**:1835-46

Weber JM and Ehrenhofer-Murray AE (2010). Design of a minimal silencer for the silent mating-type locus *HML* of *Saccharomyces cerevisiae*. *Nucleic Acids Res* **38(22)**:7991-8000

Ehrentraut S, Weber JM, Dybowski JN, Hoffmann D, Ehrenhofer-Murray AE (2010). Rpd3-dependent boundary formation at telomeres by removal of Sir2 substrate. *Proc Natl Acad Sci* **107(12)**, 5522-7

Weber JM, Irlbacher H, Ehrenhofer-Murray AE (2008). Control of replication initiation by the Sum1/Rfm1/Hst1 histone deacetylase. *BMC Molecular Biology* **9**,100

Mayrhofer S, Weber JM und Pöggeler S (2006). Pheromones and pheromone receptors are required for proper sexual development in the homothallic ascomycete Sordaria macrospora. *Genetics* **172**, 1-13

Poster presentations

Weber JM, Irlbacher H, Ehrenhofer-Murray AE (2009) Regulation of replication initiation and *HML* silencing in *Saccharomyces cerevisiae*. „*8. Forschungstag der Medizinischen Fakultät*", *Universitätsklinikum Essen, Essen*

Weber JM, Irlbacher H, Ehrenhofer-Murray AE (2009) Control of replication initiation by the Sum1/Rfm1/Hst1 histone deacetylase. „*EMBO Conference Series on Chromatin and Epigenetics*", *European Molecular Biology Laboratory, Heidelberg*

Weber JM, Irlbacher H, Ehrenhofer-Murray AE (2009) Control of replication initiation by Sum1/Rfm1/Hst1-mediated histone deacetylation. „*EMBO Workshops Gene Transcription in Yeast*", *Sant Feliu De Gioxols, Spanien*

Mayrhofer S, Weber JM, Pöggeler S (2005) Characterisation of pheromones and pheromone receptors of the homothallic ascomycete *Sordaria macrospora*. „Molecular Biology of Fungi VAAM-Symposium", *Conference Center at the St. Josef-Hospital, Bochum*

i want morebooks!

Buy your books fast and straightforward online - at one of world's fastest growing online book stores! Environmentally sound due to Print-on-Demand technologies.

Buy your books online at

www.get-morebooks.com

Kaufen Sie Ihre Bücher schnell und unkompliziert online – auf einer der am schnellsten wachsenden Buchhandelsplattformen weltweit! Dank Print-On-Demand umwelt- und ressourcenschonend produziert.

Bücher schneller online kaufen

www.morebooks.de

VDM Verlagsservicegesellschaft mbH
Heinrich-Böcking-Str. 6-8
D - 66121 Saarbrücken

Telefon: +49 681 3720 174
Telefax: +49 681 3720 1749

info@vdm-vsg.de
www.vdm-vsg.de

Printed by Books on Demand GmbH, Norderstedt / Germany